Developmental Education Preparation

Developmental Education Preparation

How Faculty Preparation in Higher Education Can Lead to Student Success

Ajai Cribbs Simmons

ROWMAN & LITTLEFIELD
Lanham • Boulder • New York • London

Published by Rowman & Littlefield
An imprint of The Rowman & Littlefield Publishing Group, Inc.
4501 Forbes Boulevard, Suite 200, Lanham, Maryland 20706
www.rowman.com

86-90 Paul Street, London EC2A 4NE, United Kingdom

Copyright © 2023 by Ajai Cribbs Simmons

All rights reserved. No part of this book may be reproduced in any form or by any electronic or mechanical means, including information storage and retrieval systems, without written permission from the publisher, except by a reviewer who may quote passages in a review.

British Library Cataloguing in Publication Information Available

Library of Congress Cataloging-in-Publication Data

Names: Simmons, Ajai Cribbs, 1981– author.
Title: Developmental education preparation : how faculty preparation in higher education can lead to student success / Ajai Cribbs Simmons.
Description: Lanham : Rowman & Littlefield, [2023] | Includes bibliographical references. | Summary: "This book will provide support for faculty across the country and help enhance their readiness for teaching developmental and corequisite mathematics courses"—Provided by publisher.
Identifiers: LCCN 2022030896 (print) | LCCN 2022030897 (ebook) | ISBN 9781475866278 (cloth) | ISBN 9781475866285 (paperback) | ISBN 9781475866292 (epub)
Subjects: LCSH: Mathematics—Remedial teaching. | Mathematics—Study and teaching (Higher)
Classification: LCC QA11.2 .S546 2023 (print) | LCC QA11.2 (ebook) | DDC 510.71—dc23/eng20221013
LC record available at https://lccn.loc.gov/2022030896
LC ebook record available at https://lccn.loc.gov/2022030897

To my family. To my husband, Al Simmons, Jr., for being with me on the journey, for support, and for sacrificing the time to keep our family functioning. To our parents, for being there to help us with the children on my journey and giving us those much-needed breaks. And finally, to my children, Rhea and Tre, for being my inspiration and motivation. I love you all.

Contents

Acknowledgments		ix
Introduction: What Do We Know?		xi
1	Developmental Mathematics Education	1
2	Developmental and Corequisite Pedagogy	13
3	Listening to the Voices of Faculty of Developmental Mathematics	21
4	Successful Professional Development Options for Community Colleges	53
5	Conclusion	71
References		81
About the Author and Contributors		87

Acknowledgments

I would like to thank my committee chair and my cochair, Dr. Mary Margaret Capraro and Dr. Robert Capraro, respectively, for their guidance and support through this research. To Dr. Jennifer Whitfield and Dr. Jamaal Young, thank you for letting me discuss my research and motivating me to keep going. I would like to thank my entire committee for pushing me to reach my goal of completing this book. Thank you for every edit, every advice, every moment of transparency in this journey.

I would like to thank my qualitative team, Dr. Tasha Bennett and Ms. Brittany Garcia-Pi, for helping me with understanding of the data. I will forever be grateful for your willingness to share time with me to complete my research and graduate. Cheers to you both.

I also appreciate every single person from the department who helped me in my seven years at TAMU—you all have been amazing to me.

Introduction
What Do We Know?

Developmental education, corequisites courses, student success, and professional development are words constantly echoed in higher education, particularly at community colleges. Ultimately, the administration assumes faculty have the skill set to achieve high success rates with students who are not college-ready. The concern with professors of higher education is that the focus of developing and acquiring strong instructional skills to combat this challenge is not a priority for administration. Most higher education mathematics faculty's undergraduate and graduate training has focused on developing strong knowledge of mathematical content. Faculty jobs in higher education require a mathematics degree, but a certificate or degree in education is not required.

Successfully teaching developmental courses requires an understanding of the background and current context in which the students are living. Many community college professors are teaching remedial mathematics to adults, which requires a different skill set than teaching to traditional college-ready students. Adding to the problem is the concept of corequisite courses. Corequisite courses combine the requirements of developmental mathematics and college-level mathematics into one classroom, oftentimes doubling the challenge of teaching these courses.

Faculty across the country understand the importance of courses that prepare students to be college-ready because we know that many students are simply not ready to step into a college-level mathematics course. Many faculty members can relate to the analogy of a large boulder, too large for a person to move, being dropped on their desk and being asked to create a diamond in 16 weeks, just as in the quote from Ellerbe (2015). Faculty members can understand the pressure of teaching underprepared students and they relate that to taking weekly whacks against the boulder; some weeks are successful while others fail completely.

Faculty often question whether they even have the correct skills to teach these students. There are times when faculty wonder if the proper training and professional development was provided to them to teach these students and transform the boulder into a diamond. Faculty members may feel the pressure to survive and just do the best they can with the skill set they do have.

Regardless of their level of training, we are confident that faculty members assigned to teach students who need to further develop their mathematical skills want their students to be successful; we want them to be college-ready. The journey toward helping underdeveloped students to become college-ready starts with faculty being equipped with the proper skills and tools to transform their boulder into a diamond so that their students can experience success in their college-level math classes. This all boils down to the notion that *college readiness for students starts with professor readiness.*

Studies have shown that remedial courses are one of the largest barriers students face in their college journey, and this barrier could almost stop their college career altogether (Logue et al., 2019). Today, it is shown that 67% of students test into developmental courses (Smith, 2016). Of those students who are placed in one or more developmental courses, only about 50% are able to complete the developmental sequence to move on to college-level courses (Chen & Simone, 2016). In Texas, 62% of those students who are deemed not ready for college and take developmental courses do not complete college-level work within a year (Morgan & Morales-Vale, 2019).

Many of the huge number of students taking developmental courses are from underrepresented groups, such as racial minorities and financially unstable students, and it is frequently those individuals who get stuck in the developmental world (Logue et al., 2019). Even though the idea of developmental education has been a norm, significant changes have been needed for some time in order to increase student success and push students to graduation, especially in Texas.

This book suggests some faculty development that can be used for teaching developmental education and corequisite courses. Complete with techniques, pedagogy, and instructional skills, when combined all together, this book can help with fostering meaningful professional development on any campus across the nation. The interviews conducted and analyzed reveal common trends in needs and characteristics of corequisite courses. Based on the themes found, professional development is suggested to aid in helping shift any negative components of those themes to help better understand the needs of teaching these unique and challenging courses. Understanding the needs of faculty and students can help to create a professional development plan that will enhance the developmental-level mathematics courses in higher education.

Chapter One

Developmental Mathematics Education

A lot of work goes into a developmental math student. It's just like someone dropped a boulder right here on my desk. They want you to produce a diamond. I take the first chip at it. A lot of times, you only get two or three whacks before they move on. After that [first course], they are nowhere near a diamond, but it's not just a boulder anymore.

—Ellerbe, 2015, p. 394

REFORM FOR DEVELOPMENTAL MATHEMATICS COURSES

Developmental education changes in Texas have been in progress over the past years to help improve the success of remedial courses. Community college students who are deemed underprepared for college-level courses may find themselves stuck in a series of developmental courses to get them prepared for college. Studies have shown that students who take developmental courses have a lower graduation rate, especially if they need to take multiple subjects of remedial courses (Scrivener et al., 2018). In 2015, the state of Texas adopted a policy that required community colleges to "implement block scheduling for at least five certificate and associate degree programs in nursing, allied health, and career and technical education" (Complete College America, 2016, p. 19). That pushed all colleges to implement this change by fall 2016.

Texas jumped on the bandwagon and continued with more campaigns to find ways to drive the success rate up for struggling college students.

In 2015, the 60x30TX became the next major education plan. With 60x30TX, there were four goals that needed to be achieved by 2030: (1) 60% of Texans 25–34 will have a degree or certificate, (2) at least 500,000 students will complete a degree of some sort from an institute of higher education, (3) the degree achieved will have identified marketable skills, and lastly, (4) undergraduate debt will not exceed 60% of the graduate's first year of wages (Texas Higher Education Coordinating Board (THECB), 2019). To reach this long-time goal, investigating how to improve student success in higher education, especially with developmental courses, is the next move for Texas. Deeper understanding of the factors tied to remedial courses will help push underprepared students to college level.

Studies have shown that students placed in remedial mathematics courses could have completed a college-level course given the opportunity (Ran & Lin, 2019). If these students were granted college-level mathematics placement, it could also increase the motivation of these students so that they felt less stigmatized by developmental-level courses (Ran & Lin, 2019). Placing students in college-level classes with the inclusion of a remedial course that supports the college-level class is called the corequisite model. Corequisite (also known as mainstreaming, co-enrollment, and course pairing) is the remedial component paired directly with the college course to advance student success in the college-level course (Morgan & Morales-Vale, 2019). The idea of corequisite education could help with this placement issue and help with increasing the number of students able to finish a college-level course. Exploring the birth of the corequisite model at various colleges will help us understand why the state of Texas has decided to take the big leap and push for having developmental education be fully modeled as corequisites.

Before the new classes were called corequisite courses, they were termed Companion Courses and were the result of an experiment with eight English students who did not make the cutoff score for college-level English courses (Goudas, 2017). The students volunteered to enroll in the course anyway, along with taking another three-credit course which would teach the students the skills they needed to be successful in completing their college-level courses. The data showed successful results and credit was awarded to the Accelerated Learning Program's founder, Peter Adams, for its creation (Goudas, 2017). Companion Courses turned into the Corequisite Model and now colleges and universities are trying models based on the increasingly positive results that are being reported.

Highlighting various colleges across the country that have been proactive in changing their developmental mathematics courses, the changes have shown documented success. One of the programs found in this search was implemented at Cleveland State Community College in Tennessee in 2008 (Squires et al., 2009). The redesigned program involved combining developmental and college-level mathematics courses and creating modules. The students complete 10–12 modules, spending one hour a week in a classroom and two hours a week in the lab (Squires et al., 2009). The results show that, before the program, the success rates were 54% and 65% for developmental and college math, respectively. After the program, the success rates were 72% and 74% for developmental and college math, respectively (Squires et al., 2009). As a side note, the program also saved the college 10% in costs, which is an added benefit.

At Santa Monica College, a research report was created to show the influence that various term lengths have on success rates at the college. For the traditional 16-week courses, the success rate was 55%. For the eight-week courses, the success rate was 61%. For the six-week courses, the success rate was 67%. The shorter the course length, the higher the success rate, and the lower the withdrawal rate. The withdrawal rate was 24% (16-week), 21% (8-week), and 17% (6-week), showing an added benefit for the accelerated learning (Geltner & Logan, 2001).

In Indiana, an analysis was conducted at Ivy Tech for 23 colleges across the state (Edgecombe, 2011). The study consisted of accelerated programs created in the 2007–2008 academic year. At Evansville campus, the campus created eight-week courses, so two courses could be completed in one semester. The traditional 16-week developmental course had a success rate of 52% for completing the low and middle level of developmental courses in two semesters; but the eight-week format had a success rate of 71% with completing the low and middle level of the developmental courses in one semester (Edgecombe, 2011). This study showed a success with this program.

South Texas College (STC) has made some modifications to their program as well. At this college, a self-paced module was created for the mid-level developmental mathematics courses. The program had 84% of the students enrolled successfully completed, in comparison to the traditional mid-level mathematics course, which only 45% successfully completed (Edgecombe, 2011). The program was conducted twice because the first set had a small sample size. Even the summer course showed a slight improvement. The traditional summer course had a success rate of 71%, but the self-paced module had a success rate of 88% (Edgecombe, 2011).

Different types of acceleration programs that are worth mentioning—but not necessarily the traditional developmental courses of pre-algebra, beginning algebra, and intermediate algebra—are the Statpath in Pittsburg and the Accelerated Learning Program (ALP) in Baltimore. Statpath is another accelerated developmental statistics path created at Los Medanos College in Pittsburg, California. It is a program where the course will teach you the elements needed to be successful in the college-level statistics class in one semester (Hern, 2012). The traditional statistics students had a success rate of 5% while the Statpath rate increased to 38% (Edgecombe, 2011).

Another successful program that does not involve mathematics but involved great improvements was the ALP at the Community College of Baltimore County. The program involved the acceleration of English courses (Jaggars et al., 2014). The success rate for the six- and nine-weeks courses was 76% and 87%, respectively. However, the traditional course had a success rate of 57%, showing a significant improvement (Jaggars et al., 2014). A combination of all the results using percentage gains on all the above programs are displayed in table 1.1 for reference.

Does the accelerated, compressed, and corequisite format work? So far, yes. The evidence is there to show the success of pushing students through challenging courses. Not only are the numbers showing greater success, but there are also added benefits, such as lower withdrawal rates and lower costs. If schools are mandated into a situation where something is not working, it can only benefit them to at least try a new method, and to try that method with enthusiasm for the purpose of fostering student success. Join a committee and find a way to make this change work for the department, make notes on what is not working with the change, and try to determine what can be done to alleviate any

Table 1.1. Success Rates of Accelerated and Compressed Community College Programs Compared to Traditional 16-Week Courses

Location of Intervention	Year	Before Intervention	After Intervention
Cleveland State Community College	2009		
Developmental Math		54%	72%
College Algebra		65%	74%
Southern California Community College	2010		
6-Week Course		57%	76%
8-Week Course		57%	87%
Ivy Tech Institutes of Indiana	2009		
Accelerated Courses		52%	71%
Reading		25%	58%
Los Medanos College	2009		
StatPath		5%	38%
South Texas College	2010		
Self-Paced Semester		45%	84%
Summer Course		71%	88%
Santa Monica College	2001		
6-Week Course		55%	67%
8-Week Course		55%	61%
Community College of Baltimore	2015		
6-Week English Class		57%	76%
9-Week English Class		57%	87%

problems. Even if not everyone agrees with the change, everyone can at least agree to try for the sake of student success. Acceleration can be a step in the right direction with everyone's help in implementing the change.

After investigating the accelerated and corequisite models in states in numerous areas, it was possible to see that Texas was next in line for implementing change. "We took a look at the models around the country and saw a number of states where the corequisite models are working better than the other jumble of developmental education models," said Raymund Paredes, commissioner of higher education for the Texas Higher Education Coordinating Board (Smith, 2017). Along with colleges and universities across the country, institutions in Texas are discussing corequisite education to accelerate students from developmental mathematics courses to successfully complete college-level mathematics courses.

Developmental courses are the prerequisite for most college courses where students are deemed underprepared via standardized placement tests (Koch et al., 2012). According to a study on accelerated programs, almost 60% of community college students are referred to one, two, or three levels of developmental-level classes (Jaggars et al., 2014). "Developmental Courses in Texas are primarily composed of incremental courses designed to bring underprepared students to the level of skill competency expected to entering college freshmen" (Booth et al., 2014, p. 2), in other words, making them college-ready. Developmental education in Texas has been evolving over the past years to help improve the success of remedial courses.

Community college students who are labeled as underprepared for college-level courses may find themselves stuck in a series of developmental courses in order to prepare them for college. Even if a student only has one prerequisite developmental course, they could possibly find themselves not successfully completing a college-level course (Morgan & Morales-Vale, 2019). Data have shown that only about 5% of students needing three levels of developmental courses will even make it to a college-level course, let alone pass it (Adams, n.d.). The solution that has been brought forth is the corequisite model to assist in increasing the success of students currently qualifying for developmental courses, especially with mathematics.

Massachusetts was one of the first states to adopt and implement some form of corequisite courses. At the Community College of Baltimore County (CCBC), they created an Accelerated Learning Program (ALP) that mimicked the model of corequisite (Emblom-Callahan et al., 2019). The data revealed that students displayed success with the ALP program and this motivated other colleges to implement this model on their campuses, and, as a result, more professional development was provided to help advance this model (Emblom-Callahan et al., 2019).

Tennessee followed as another one of the initial states that backed the plan for corequisite courses for developmental education students. Tennessee also pushed for a statewide initiative. All of their 13 public community colleges created corequisite models to determine if this new model had positive effects and could help the students of developmental courses in Tennessee (Complete College America, 2016). With the traditional model of remedial courses in 2012, only about 12% of students were able to complete their remedial courses and complete a college-level course as well (Ran et al., 2019). With the implementation of corequisite courses in Tennessee in the fall of 2015, the data shows that about 51% of students in those community colleges were able to pass a college-level course (Ran & Lin, 2019).

No matter the ACT score of the student, data shows that, with the corequisite models, students are demonstrating significant improvements in passing college-level courses in contrast to the traditional prerequisite model studied in fall 2012 (Smith, 2017). Though other states such as Florida and California had trials of corequisite education as well, the data collected from Tennessee had an impact on the push for other states when examining corequisite education models.

Texas was the next state to implement changes in developmental education. "We took a look at the models around the country and saw a number of states where the corequisite models were working better than the other jumble of developmental education models," said Raymund Paredes, commissioner of higher education for the Texas Higher Education Coordinating Board (Smith, 2017, para. 4). Research shows that in Tennessee, only about 3% of students scoring less than 13 on their ACT passed a college-level course. When enrolled in the corequisite model, about 58% of students in the same category were able to pass college-level courses (60x30TX, 2018). This is a significant improvement and statistics like this led to the push to follow the Tennessee model in the state of Texas.

The shift to corequisite education began across Texas in June 2017, when governor Greg Abbott signed a mandate, House Bill 2223, explaining that any student enrolled in a developmental course must also be enrolled in its college-level course counterpart (House Research Organization, 2017). Texas was one the first states to push for such a huge reform statewide and actually pass a house bill to accomplish the mission (Morgan & Morales-Vale, 2019). Currently, 75% of developmental courses at any college or university in Texas must be of the corequisite model. A timeline was also set up to help with this transition.

Texas was placed on a timeline to start full participation in corequisite education. House Research Organization (2017) stated that in Texas, the state's public colleges and universities with developmental courses had until 2018 to get 25% of their developmental courses paired with college-level courses to create corequisites (60x30TX, 2018). To continue the transition, the state wanted 50% of the developmental courses paired by 2019 and 75% paired by 2020 (60x30TX, 2018).

Why the gradual transition? The board wanted to give colleges and universities a chance to fight back with a better alternative. The gradual change allowed campuses the freedom to come up with a more successful plan for their campus, allowing them to alter the timeline or push for the corequisite (Smith, 2017). Some campuses have tried alternative plans for implementing the corequisites model, but very few campuses took advantage of attempting alternate plans for developmental education outside of corequisites. Thus, the timeline has basically remained as originally designed for developmental education.

PROS AND CONS OF COREQUISITE STRUCTURE

As the corequisite model is implemented into all colleges and universities in the state of Texas, there are reports of support and initiatives to continue the model's progress. The Texas Success Center (TSC) had a Texas Corequisite Project webinar this year that shared some content and pedagogical techniques for teachers teaching in corequisite education classes. A proactive approach has been provided for colleges and universities in order to allow them the freedom to attempt different techniques for their particular campuses. Paredes stated that the state is "not suggesting we all do the corequisite education in Texas the same way" (Smith, 2017, para. 3). Not mandating specifics was the most effective because it allows for different techniques and approaches to be implemented that complement different faculty groups. This process also allows faculty to come together and share their successes and failures with this novel corequisite teaching.

A webinar held by TSC allows colleges and universities to come together to share their experiences with implementing corequisite models. The webinar included representatives from the campuses of Texas A&M University–Kingsville, College of the Mainland, University of Texas Rio Grande Valley, Amarillo College, North Lake College, Victoria College, and Prairie View A&M University (Texas Success Center, 2020). In this one-day webinar, all of these stakeholders were afforded opportunities to share their successes and offer suggestions for implementing corequisites.

Another benefit of enrollment in corequisites is the ability to provide instruction for students who took a placement test and scored below level. Placement in a credit-bearing college course can help the morale of those students who believe they are college-ready but fall short with placement test scoring. It has been found that 86% of students believe they are ready for college academically; however, 67% of these students are placed into remedial courses, showing they are underprepared for college-level courses (Ran & Lin, 2019). A study conducted to analyze standardized placement tests such as ACCUPLACER and COMPASS shows that 25% or 34%, respectively, of students are severely misassigned to college courses (Scott-Clayton et al., 2014). Thus, students may be placed in classes they are not prepared for (over-placed) and vice versa; there may be situations where students are not allowed to take courses they are ready to take (under-placed).

Placement tests, such as the Texas Success Initiative Assessment, can place students in remedial classes even though they may be college-ready (Logue et al., 2019). Corequisites place students in college-level classes, boosting their overall morale and allowing them to be college students in credit-bearing courses (Logue et al., 2019). Misplacement due to standardized testing can be eliminated by fully implementing the corequisite model. Some states have totally eliminated standardized assessment once they fully establish the corequisite model (Adams, n.d.). Thus, the lack of misplacement of students in remedial courses and an improvement in the morale of students is a definite benefit of corequisite models.

The transition to corequisites remains a challenge. Some campuses have resisted the change and present some pushbacks. One aspect that has been a challenge for some campuses is the need for qualified instructors. Instructors who teach developmental courses may not be qualified to teach college-level courses, which usually require 18 credits of graduate-level mathematics for most schools (Complete

College America, 2016). Even those instructors who are college-level qualified may not have the training to teach using effective developmental pedagogy (Smith, 2017). This leads to some campuses having corequisite courses taught by two different instructors: one to fulfill the developmental side and one to fulfill the college-level side. Whether the corequisite is taught by one instructor or two, both methods have shown greater success than the traditional prerequisite format; however, instructor collaboration involves more time and preparation from both instructors to improve the students' learning.

Cost is an important factor for students and an aspect that campuses consider when it comes to students. The corequisite model has been advertised as saving more money for students in comparison to the traditional prerequisite model. This is only because students are less likely to enroll in a large number of remedial courses without earning college credit. The initial cost for students is more per student when they enrolled through the corequisite pathway (Smith, 2016). This new pathway requires students to pay for two math courses in one semester, rather than one. Unfortunately, if they do not pass either of the courses, this can end up costing more. However, data has shown that the cost is lower for students overall due to not having to take a series of remedial courses (Complete College America, 2016).

The differences in how each campus delivers its corequisite courses could lead to confusion on how to implement the model for Texas colleges and universities. There is not an empirical study done yet to determine which method has resulted in the most effective success rates. Thus, this can lead to questions like, "if one math corequisite course on one campus saw better results than another college, what are the differences between the way those courses were delivered?" (Smith, 2016, para. 12). The data being presented would help campuses in considering better options for their campuses. Presently, because the corequisite model has so recently been implemented, there are few initiatives requiring training. Thus, most campuses are just using a trial-and-error approach, employing various strategies with no real training on how to track the data for improvements.

In Texas, all colleges and universities should be at least at the 75% level of corequisite model with their developmental courses. Exploration is being initiated for professional development. There are some forms of statewide professional development available, such as the

Texas Corequisite Project organized by Austin Community College. This project offers statewide professional development for colleges and universities to help improve the success rate with implementing the corequisite model by first understanding the campus through surveys (Austin Community College, n.d.). This project is not a requirement for colleges; only the implementation of a corequisite model is required. However, there is value in professional development and thus this training should be required for instructors to better understand the pedagogy for developmental teaching.

Campuses outside of Texas have had success when they mandate training. One example is Tulsa Community College (TCC) which requires faculty to attend summer training and offers stipends to adjuncts for their time spent attending the training (Tulsa Community College, n.d.). Overall, the success of corequisite implementation can continue even without the proper training, but professional development would be an added beneficial component to the plans in Texas for faculty to better understand the corequisites and best practices in classroom instruction for these unique students.

Professional development is an opportunity for faculty to improve in many areas, such as content knowledge, pedagogy, and dispositions (Burrows, 2015). The word pedagogy may not be a common word used in college-level courses; however, exploring the various forms of pedagogy that can be used in the classroom can definitely help faculty members develop new material and approaches to teaching it (Burrows, 2015). Research shows that pedagogical learning is more effective for students, so training faculty helps with improving their pedagogical skills, hopefully increasing the success rates for students. How this training is conducted is very important to making sure the faculty learns about the most effective enhancements for their teaching. The next chapter explores some of the pedagogical techniques utilized on campuses with teaching college-level students.

Chapter Two

Developmental and Corequisite Pedagogy

It seems reasonable to ask, were faculty more attuned to the impact of their pedagogical choices, might they become more agile and adaptive in the classroom and might this have a significant impact on student persistence and completion?

—Mellow et al., 2015, p. 18

WHAT ARE WE DOING?

A community college can be referred to as a teaching college—a place where emphasis is placed on the instruction in the classroom and less on the research outside of it. Though there have been some critics who wonder if community colleges are a "poor infrastructure and weak culture for supporting professional development about instruction," there are still a variety of pedagogical practices that have been proven successful for faculty and students' success (Edwards et al., 2015). Around nine million students attended a community college in the 2016–2017 year, and that number made up 39% of undergraduate students (Weiss & Headlam, 2019).

With such a significant number of students attending community college or seeking to become college-ready, the preparation needed for faculty to be ready to teach this population becomes extremely important. Close to 60% of community college students need to take a remedial mathematics course, which indicates that they are not ready for college-level mathematics (Ngo, 2019). There are colleges and universities across the United States that have incorporated some type of alternative to the traditional developmental courses offered for mathematics. Among those alternatives are different forms of accelerated programs, including corequisites that are aimed at pushing underprepared students to enroll in college-level courses. Many colleges and universities have developed their own variations of the corequisite model. This chapter examines the collection of studies that explore instructional skills and practices as well as effective teaching and learning pedagogies utilized by professors at two-year institutions, especially those that teach developmental mathematics courses to the many students that need them.

Examining different studies involving various campuses will help in creating a collection of learning pedagogies which, when enacted effectively in classrooms, have been proven to be successful among students. Two-year institutions of higher learning have already been proven to be the pit stop for many college students seeking to become college-ready, and faculty members need to match that same desire by equipping themselves with a tool chest of strategies that are most effective with adult learning pedagogy (Ngo, 2019).

An interesting observation is that developmental education is compared to the learning outcomes of classes in secondary education, but those who teach secondary education are required to have certification of their ability to teach. Though this same teaching certification is not required for a college instructor, techniques learned through teaching certification programs could prove beneficial for instructors of adults, possibly even more than for those in secondary education. Some studies have shown success with pedagogical teaching strategies in higher education classrooms; however, many colleges do not require teaching education. Success in developing effective pedagogies with adult learners could prove beneficial in improving academic success for community college students.

A beneficial benchmark for checking for mathematical proficiency in the classroom can be encompassed in five components put together by the Mathematics Learning Study Committee under the National Research Council (Cox, 2015). Those five components are the following: conceptual understanding, procedural fluency, strategic competence, adaptive reasoning, and productive disposition (Cox, 2015). Along with the idea of these components, the American Mathematical Association of Two-Year Colleges (AMATYC) has a set of "pedagogical standards that balances students' acquisition of procedural skills and conceptual understanding" (Cox, 2015, p. 267). AMATYC created a document called *Beyond Crossroads*, and it is the second document created that proposed a set of "implementation standards" to highlight teaching areas like learning environment, program development, and instruction (Blair, 2006, p. 1). This document was created to be a "standards-based mathematics education that promotes continuous professional growth" (Blair, 2006, p. 2).

The standards specifically focused on pedagogy are the following: teaching with technology, active and interactive learning, making connections, using multiple strategies, and experiencing mathematics (Blair, 2006). These standards can be useful with making sure the instructor is striving to reach the peak teaching potential to benefit the students.

How much are these standards of pedagogy stressed in the college world? The answer is not completely clear; however, let us think about how many instructors are actively including these standards in their preparation, semester to semester. Conversations are held about the content that is delivered, but few discussions are conducted surrounding how to deliver this content. Looking further into schools to see which of these standards are being used to fulfill the practice of pedagogy, one can see the power of what the community colleges already have in place. Standards are created to reach a wider range of students with some of the best and proven techniques in the classroom. The purpose of standards is to place faculty on the same path of pedagogical goals in teaching. These standards should be emphasized and required in the two-year college world.

Making connections in the classroom can be marked as the most important component in a mathematics classroom. Connections build trust. One way to build trust in a classroom that can have anywhere from 25–60 students is with a supplemental instructor (SI) program. At LaGuardia Community College, there is an SI program called the Academic Peer Instruction (API) program which they are using to support the corequisite models on their campus (Jaafar et al., 2021). The researchers found that students who took developmental mathematics courses with an API had a 70.95% pass rate, which compared to the 57.38% pass rate for those who did not have an API (Jaafar et al., 2021).

These rates where increased the more API sessions a student attended outside of class; almost 100% of the students who attended 10 or more sessions passed the course (Jaafar et al., 2021). The components that make API so valuable and impactful range from understanding learning styles to collaboration. "Knowing learning styles creates a foundation of effective learning" (Jaafar et al., 2021, p. 40). The collaboration referred to the connections made with other peer programs; in this connection, a culture is created "where students do not feel ashamed about seeking help" (Jaafar et al., 2021, p. 40). API programs contribute to the growth of classroom connections, thus increasing students' success.

Alternative methods on how to teach developmental mathematics continue to be a work in progress over the years. One alternative that is popular with mathematics departments around the world is the adoption of computer programs to help students. Popular programs are MyMathLab, WebAssign, Connect Math, or ALEKS. To add onto the list, there is an institution of higher education in Texas, Tarrant County College (TCC), that created a self-paced, computer-assisted developmental mathematics course sequence to help students who are not college-ready (Weiss & Headlam, 2019). The program is called ModMath and contains six five-week modules to assist students in navigating through developmental math topics, which is delivered in MyMathLab (Weiss & Headlam, 2019).

This program is a way to keep the computer-based assignments but offers more structure with modules. Components include a diagnostic assessment, modularized courses, computer-based instruction, and on-demand assistance. The program addresses the pedagogical challenge in developmental math which is "that the courses typically serve heterogenous students with a wide range of academic abilities, learning styles, and personal needs" (Weiss & Headlam, 2019, p. 488). Therefore, if a teacher is not equipped with the ability to address every type of student who is enrolled in their classes, the ModMath program could allow students to work on their lack of prior knowledge skills. As much as teachers may prepare, there may be areas where they could use assistance—this program was designed to help with that. However, does this computerized pedagogy design work and is it showing success? The answer is yes; the results are detailed below.

The results are directly related to characteristics that increase student success. Surveys from students in this ModMath program show that

68% of students felt their instructor spent a considerable amount of time individually with students in comparison to the 32% of students in the control group (Weiss & Headlam, 2019). Students also noted that working alone on math problems was incorporated more in the ModMath course, with 81% of students agreeing that this was beneficial in the ModMath course versus 54% agreeing in the control course (Weiss & Headlam, 2019). There were positive results for less lecture time, leaving more time for hands-on learning. These three components can positively contribute to student success in one way or another. This computer-based course adds to the list of pedagogical techniques.

Student-centered learning is noted as having a stronger impact than a faculty-centered classroom. Ivy Tech Community College (ITCC) has a corequisite model that was fully implemented by fall 2014, and there are components that make the program a successful addition to their curriculum (Royer & Baker, 2018). The corequisites are equipped with peer-led groups that are facilitated by faculty members; students are able to carry on conceptual conversations about lessons while faculty guide with leading questions (Royer & Baker, 2018).

Student-centered learning solidifies the knowledge gained and raises the level of learning to an expert understanding. One of the most significant student success challenges is the presence of academic shame, and conversation is a tool that can eliminate that shame among students (Royer & Baker, 2018). The corequisite model at ITCC was based on part of the new Mathway, which combines developmental and gateway course learning (Royer & Baker, 2018). Combining methods is key to creating a course that benefits the students in the long run as it is directly related to preparing students to be college-ready.

Combining methods adds to the overall value of a student's college education. Bloom's taxonomy framework has been used and studied by teachers since 1956 in hope of understanding the levels of how people learn and what the most valuable level of learning is (Heick, 2012). It is easy to get caught in the cycle of memorization in mathematics because the field is covered in rules, formulas, and equations that apply those rules. Sometimes the value of understanding is shaded by the simple thought of having the students' know the math rules. Students can evaluate how professors teach their courses and what seems to be the most valuable lesson in terms of understanding.

In the study by Cox (2015), two colleges were observed. Most classes had the traditional teaching scenarios where the instructor delivered the lesson, and then practice problems were provided to repeat the rules taught to the students. The instructors were clear in their teaching and one instructor even noted how important it is "to learn the rules" (Cox, 2015). Every now and then, an engaging activity was enacted in the classroom to change the pace. However, there were two instructors who attempted a different approach that was based on building on previous knowledge and conversation. In this practice, students were able to discover the lesson before any "rules" were given. These classes "offered opportunities for students to engage in meaning-making exercises and develop more conceptual understandings of mathematical principles" (Cox, 2015, p. 273).

The power was given up from the teacher to the student to blend their previous knowledge with the discovery of new practices in mathematics. The instructors of these courses would ask leading questions as simple as "What does this mean [to you]?" (Cox, 2015, p. 273). Deeper levels of learning have proven to help students retain knowledge better. Pedagogy on how to guide the students through a conceptual conversation on mathematical ideas can add value to the classroom and break the routine of learning the rules, practicing the rules, and repeating the process.

At LaGuardia Community College, some instructors incorporated a strategy with themes and tags to help with pedagogy in the mathematics classroom; the concept of themes and tags was part of a professional development program, "Taking College Teaching Seriously," designed to enhance teaching and learning (Khoule et al., 2015). The list of tags provided were the following: caring, inclusiveness, differentiated instruction, multimodal instruction, contextualization, collaboration, adaptability, higher-order thinking, self-reflection, high expectations, structured lessons, time on task, connections, scaffolding, and assessment (Khoule et al., 2015).

Using these tags, a faculty member can highlight areas in their current lesson that can be strengthened with pedagogical practices. Another bonus to this professional development is the sense of community that is built among those participating in the program, where ideas and discussion are shared about their class experiences (Khoule et al., 2015). The collaborative learning environment aids in professional growth, shifting to a more pedagogical approach in teaching college students.

To further discuss how delivery of instruction matters, research was conducted in North Carolina after the redesign of developmental courses. The key redesign components were the change to four-week classes and incorporating deeper understanding of material by having multiple representations of the material used to teach mathematical concepts (Bishop et al., 2018). Instructional techniques were provided; however, instructional delivery was the choice of the college, between computer-centered, teacher-centered, or student-centered (Bishop et al., 2018).

The results of this study show a significant difference in success rates in favor of those who engage in student-centered learning, which is also commonly reported by many instructors as a strategy that can make a difference. The success rate of students involved in student-centered (67%) versus teacher-centered (58%) instruction demonstrated that a student-centered methodological strategy enables more students to be successful. When instructors employ student-centered pedagogical strategies, students are provided opportunities and time to process the content and have their moments of productive struggle with the material, enabling them to grasp concepts more deeply (Bishop et al., 2018).

One of the students in the student-centered classroom noted that "the responsibility of learning belongs to everyone in the classroom, not just the instructor, and this single strategy can keep the students engaged in their success in the course" (Bishop et al., 2018, p. 713). The classroom experience holds a tremendous amount of influence on the impact mathematics can have on a student and whether they move forward in their academic careers.

Instructors may have students who share the same error patterns, and having conversations about their errors can help students incorporate some of their ideas and gain understanding surrounding their misconceptions. Helping students develop number sense allows them to understand the big idea around important mathematical topics. Developing number sense allows students to make connections and empowers them to estimate. Estimation is also helpful when students use inductive reasoning. Another helpful way to develop inductive reasoning is to employ an interview procedure with each student so they can talk through any problems which they have solved incorrectly.

Conducting an interview can allow students to verbally discuss what they are thinking when they attempt to solve algebra problems. These conversations can help a student learn to think aloud and verbalize their reasoning as they engage in solving problems (Joyce et al., 2015). This should be a routine part of instruction for every developmental educator instructor. It is necessary for us to teach our students how to think when solving problems. Joyce et al. (2015) suggest that we cannot merely ask our students questions and assume that they will know how to get an answer. Students need to be taught how to analyze and create. Analyzing and creating should not be a one-day instructional strategy in the classroom; these processes should be ongoing in order to assist students in gaining these skills independently.

Inductive reasoning can be utilized as a component of professional development for instructors of developmental mathematics courses. It has been proven that professional development is an additional step in teaching at community colleges that actually promotes improvement in student achievement in developmental courses. Implementing some of the suggestions in chapter 4 can lead to critical and valuable additions to improving instruction, thus helping our students to improve their grasp of important mathematical concepts. Having forums where instructors can share activities that help students comprehend and develop their self-discovered knowledge enables instructors to make improvements to their future courses.

The next chapter is about hearing the voices of some instructors who have or are currently teaching the developmental model on their campus. On a quest to find any common themes among the professors, the similarities and differences can open our eyes to ways to achieve change in our developmental courses for the possibility of greater student success. Focused faculty development can be considerably helpful for campuses. Many campuses have some sort of professional development for their instructors; some campuses may even require it.

Having an option for mathematics professors that can reach into directly improving the unique characteristic of a corequisite classroom would be a welcome addition to a campus. No training is "one-size-fits-all," but a variety of options that have been proven to help is of benefit to mathematics departments as they search for what works for their students. Let us hear the voices of some instructors and collect the themes that emerge.

Chapter Three

Listening to the Voices of Faculty of Developmental Mathematics

> Just think about it, why would you become a doctor? To help people that are sick, you know, not people that are well. . . . Become a math teacher to help people who struggle with math.
>
> —Interviewee from study

WHAT DO WE SAY?

Developmental education, corequisite courses, student success, and professional development are important concepts in higher education, particularly at the community college level. The difficult task for higher education faculty is to determine how to apply them meaningfully in their interactions with students through development of their teaching pedagogies.

This presents a challenge, given that 68% of community college students are testing at a mathematical level that requires them to take at least one developmental course (Royer & Baker, 2018). Successfully teaching developmental courses requires an understanding of the background and current context in which the students are living. Successful teaching requires an understanding of learning styles and mathematics pedagogy that embraces adults' ways of learning. Many community college professors are teaching remedial mathematics to adults, which requires a different skill set than teaching to traditional college-ready students; adding to the problem is the concept of corequisite courses.

Corequisite courses combine the requirements of developmental mathematics and college-level mathematics into one classroom, oftentimes doubling the challenge of teaching these courses.

The aim of this book is to create a blueprint that can be used for teaching developmental education and corequisite courses, complete with techniques, pedagogy, and instructional skills to help guide meaningful professional development on any campus across the nation. Being a document for both scholarly and novice readers, it supports the reform that most states are witnessing with corequisites. In addition, it contains current research that is based on practicing professors' perceptions and approaches to teaching developmental and corequisite classes. One step toward pulling these resources together is to hear the voices of faculty who teach developmental classes, specifically corequisite courses.

Understanding the needs of these faculty and students will help to create a professional development plan that will enhance the developmental-level mathematics courses on community college campuses. The questions that really steered the journey of this book were the following:

- Are there key components that professors of developmental education share in their experience with teaching underprepared students?
- What components would be most beneficial in professional development to help assist faculty in teaching underprepared college students, specifically those taking algebraic corequisite courses?

The framework for conducting this research was based on a qualitative study done on the job satisfaction of faculty members in *A Qualitative Method for Assessing Faculty Satisfaction* (Ambrose et al., 2005). In the quest to collect feedback and data from faculty about their job satisfaction, this group opted to do semi-structured interviews in lieu of predefined surveys (Ambrose et al., 2005). The original plan for my study was to distribute Likert scale surveys to multiple faculty on various campuses. However, there was a need to utilize a more intimate approach to uncover exactly how the faculty felt as they taught developmental education students; in conducting interviews, the conversations were rich and confirmed some of the whispers about teaching developmental courses.

The semi-structured interviews also brought to light some new challenges and even successes of the corequisite model. Ambrose stated that the surveys "limit the range of possible responses from participants and isolate subjective perceptions from objective events and experiences that have shaped them" (Ambrose et al., 2005, p. 807). When collecting data from the faculty who teach corequisites, the goal was to understand the feelings of those who face the students every day; however, those voices might be limited by the predefined surveys. The feelings and perceptions of faculty members were unknown from campus to campus; interviews could shed light on opinions about the mandate for corequisites. After the interviews were complete, the coding revealed an overall universal pattern that overlapped campuses.

During the first step, three colleges were identified that have established corequisites on campus. For the sake of privacy, they were identified with the following names: College A, College B, and College C. At each college, two types of faculty members were identified who had taught corequisite mathematics courses: (1) those who had some form of education degree or teaching certification, and (2) those who had not received an education degree but had a content-specific degree (mathematics). The choice of this criterion was not only to identify common themes that might appear but also to compare faculty who have pedagogical teaching knowledge with those who do not. Are there areas where faculty without education degrees struggle more than those with degrees? The data collected revealed the answers to this important question as the results were analyzed.

Once these faculty members were chosen and the research review boards were approved to conduct the interviews, a set of questions was created to proceed with a semi-structured conversation with each participant. The following questions were asked:

1. Explain your modality of the college algebra corequisites.
2. Provide the most successful component of corequisites on your campus.
3. Provide the hardest challenge of corequisites on your campus.
4. Does your campus offer professional development to help with corequisite teaching?
5. If yes, what components of the professional development are most useful?

As we discuss the results of data, it is important that the terminology is explained and also the descriptions of the courses. The corequisite course is a course that is taken concurrently with another course. Throughout this book, the corequisite course is the developmental mathematics course (which is not college-level) and the course that the corequisite is paired with is college algebra (which is college-level). As discussed in chapter 1, in Texas, a developmental mathematics course must be paired with a college-level mathematics course.

These two courses must be taken together, in part to help students succeed in their college-level courses and increase their chances of completing college. In the interviews, the words developmental, fundamental, remedial, and corequisite were all used interchangeably, because they all are referring to the noncollege-level part of the two courses. When the interviewees were speaking of the classes being taught individually, they would use the word "standalone." In other words, before the mandate, students had to take the developmental course as a prerequisite before taking any college-level course, so they referred to it as the "developmental standalone." Interviewees referred to the college-level algebra course as the "college algebra standalone."

This simply means not in conjunction with the corequisite course. Many campuses have their own individual names for their corequisite courses but, for confidentiality, they will all be referred to as corequisite courses. It is important to clarify this for the reader as the codes are shared.

The first question, in reference to the modality of the college algebra corequisite course, was asked via email or discussed at the beginning of the interview. This was an objective question asked to obtain the details of how the classes were taught at the individual's campus.

College A had a variety of different modalities. Some classes were taught by two instructors, where one instructor taught the corequisite and the other taught the college algebra. Most classes were taught by one instructor, where the instructor taught both corequisite and college algebra. The two instructors interviewed taught both parts of their courses. College B had instructors for each part: one instructor taught the corequisite course and another instructor taught the college algebra. The students were the same in both courses—just the teacher changed. The interview included the instructors of one class set: one who taught the corequisite and the other who taught the college algebra. College C had two instructors in one class—the corequisite and college algebra were taught as one course and both instructors were present during the entire class. They basically co-taught different parts of the course together. The interview was conducted with two instructors from one of the courses.

With every college, there was one instructor who had a background in education and one who possessed a purely content-based background. The education background consisted of either an education degree or teaching certificate from teaching K–12 at some point in their career. This background was determined before the request was sent to interview to ensure that these multiple viewpoints were captured.

The interviews were recorded and then transcribed. Responses were edited because some words were not transcribed correctly from the recording. Once all the corrections were made and any details that identified the interviewee were removed, the recordings were destroyed to maintain the confidentiality of the participants. The cleaned-up transcripts were shared with a qualitative team for analysis. The group started with an individual review of the transcripts to get a feel for the tone of the respondents. The data collected from the interviews was analyzed using the grounded theory technique (Charmaz, 2000).

The goal was to find common themes among the schools by doing careful analysis of the data with line-by-line coding, creating a thematic analysis among the different institutions. This line-by-line coding is a grounded theory technique where codes are created while analyzing the data (Charmaz, 2000). The qualitative data from the interviews was analyzed first to create categories and subcategories from codes developed from the line-by-line coding (Charmaz, 2000). Using this technique ensured the voices were heard from the data responses and codes were not just based on the researchers' own personal feelings (Charmaz, 2000).

Analyzing can utilize some text-in-context coding to link qualitative and quantitative data (Creamer, 2018). Having interviews on multiple campuses provided opportunities to create validity from the developed themes and categories through triangulations by analyzing the research question from multiple perspectives (Creamer, 2018). One alteration the qualitative team discussed was how to break the data up for coding. A semi-altered change to Charmaz's method was to split the transcripts by thoughts instead of going line by line. Some lines included multiple codes; then there were times when two or three lines included one code.

This minor change helped the research team go through the data in order to make more sense of the themes that were developed. This change also helped with comparing similar codes among the three researchers.

The inductive coding allowed the group opportunities to start from scratch and create their own set of codes from the words of the participants. This inductive approach led us to conduct a thematic analysis of the results to determine what common themes were developing among the interviews. From the individual codes and subcodes created, a primary list was created combining all three sets of results. The primary list helped edit the wording of the codes and helped with repetition. After the primary list was solidified, the codes were grouped to find the themes that were emerging.

To create the themes, the codes were grouped into similar topics to act as an umbrella for several codes. To keep track of various colleges and the type of professor, the codes were color-coded to track where they came from. This was a very important piece to track in order to determine if there were any similarities between instructors with education training versus faculty members with content-only training (figure 3.1). Using an app called Post-it helped with the mobility to shift codes and allowed the coders to wrestle with which umbrella they fell under. The themes created held positive and negative viewpoints of that theme, which split the themes into two categories highlighting the upside and downside of the various themes.

After rounds of shifting codes and discussing thoughts about their placement, thirteen themes emerged that most accurately summarized the thoughts and words of the interviewees. The themes are the following: (1) Administration issues, (2) characteristics of students, (3) corequisite course characteristics, (4) faculty expectations, (5) format of course, (6) job satisfaction, (7) matriculation success, (8) course misconceptions, (9) blending courses, (10) resources, (11) standalone remarks, (12) understanding students, and (13) time. Among these thirteen themes, nine of them held a higher number of codes and were shared across all three campuses, and these codes also showed a greater influence than the other four themes.

The goal was not to quantify the qualitative results but more to highlight the areas where overlap occurred. These nine codes are further elaborated upon in the discussion that follows: (1) Blending courses, (2) characteristic of students, (3) corequisite course characteristics, (4) course misconceptions, (5) faculty expectations, (6) format of course,

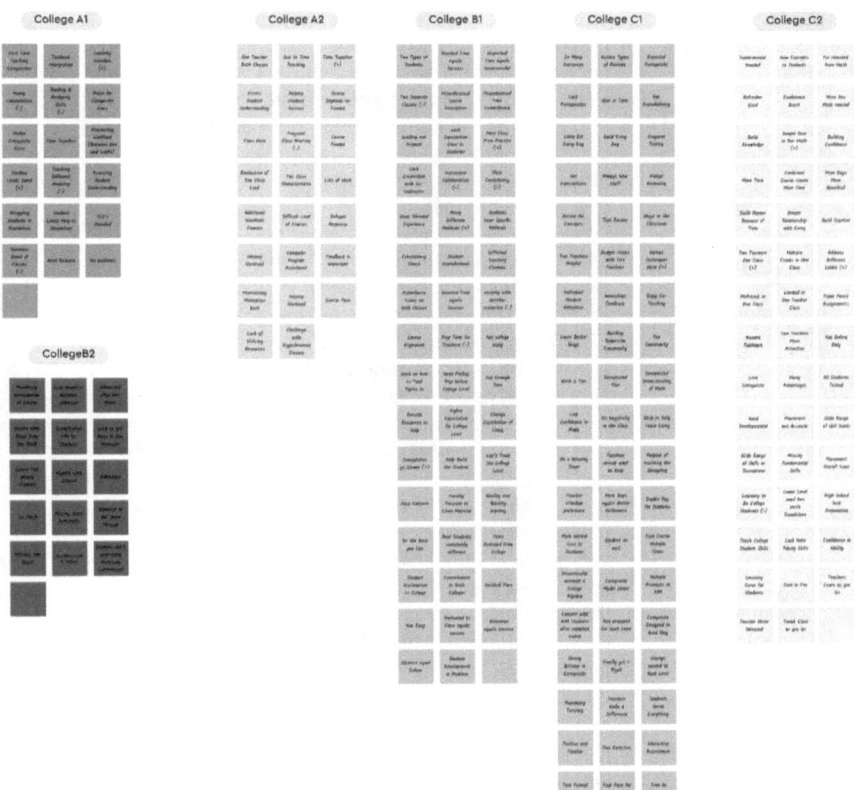

Figure 3.1. Displayed is the list of codes created from the interviews. The codes were collected and organized in Post-it Notes App. The colors are displayed to differentiate the teaching.

(7) job satisfaction, (8) time, and (9) understanding students. To further understand each theme, a definition and description is provided in tables 3.1–3.3 below.

Table 3.1.

Theme	Description
Blending Courses	A corequisite course is a course taken concurrently with another course. Therefore, a student will be enrolled in two courses in the same semester. This theme is dedicated to the comments shared concerning the blending of these two courses. Examples: "I feel like blending those two [courses] it's a really hard blending." "We can present things in different ways, and we can attend to students individually as well as in groups because of having two classes."
Characteristics of Students	Students in a corequisite course come into the course with a variety of characteristics based on feelings, skill levels, self-esteem, and background. This theme addresses the characteristics found in the students of corequisite courses. Examples: "I feel like I know what to expect, with my corequisite class, I know what level they're at, less variability." "Going back to school after all those years, not really understanding, like the amount of time it takes to be successful in college classes."
Corequisite Course Characteristics	The corequisite courses have a collection of characteristics that shed light on this unique environment. This theme has a list of some characteristics that make the course unique. Examples: "The Just in Time review is a successful component of corequisite courses." "Corequisites are an intense workload with a lot of heavy computation."
Course Misconceptions	When students are placed in a corequisite course, they may not fully understand what they are signing up for. The theme shares some thoughts that confirm some misconceptions students have about the course. Examples: "The students realize that these are two different classes . . . you do have all the these assignments . . . Sometimes I do run into the students saying this is a lot of work." "I think the students are surprised that they are enjoying the class, that they look forward to their class."

Themes were developed from interviews with faculty who teach or have taught the corequisite mathematics classes. The list includes the description of each theme.

Table 3.2.

Theme	Description
Faculty Expectations	A faculty member who teaches a corequisite course, especially for the first time, may have a set of expectations for the course. The expectations range from classroom-related to student-related. This theme shared some expectations of the faculty when it comes to corequisite courses. Examples: *"The students should feel like they're sort of this group who is making it together."* *"The most successful part of this class was that the textbook had an integrated review built in."*
Format of Course	Format of how to teach a corequisite course is not consistent across the United States. Many schools, especially in Texas, have been using a trial-and-error method of how to create the best environment for students. This theme addresses some of various formats and structures that were discovered in the interviews. Examples: *"With online courses, there may be additional kinds of factors that you wouldn't have with an in-person course."* *"I think it helps with the eight-week format, have it in a hybrid-type format."*
Job Satisfaction	When they are asked to teach a corequisite course, how does this change in developmental math affect faculty job satisfaction, and simply how do faculty feel teaching the course? This theme is all about the feelings of faculty when it comes to corequisites and their work environment. Examples: *"It's just too much and it's too much material to teach."* *"Whenever I'm working somewhere and I'm having fun, I feel like I do a better job."*

Themes were developed from interviews with faculty who teach or have taught the corequisite mathematics classes. The list includes the description of each theme.

Table 3.3.

Theme	Description
Time	Corequisite courses have this connection with time that needs to be addressed. This theme evaluates some of the comments made about time when it comes to corequisite courses. Examples: *"They have time in class to practice with their friends."* *"In a standalone, because there is so much to cover, we can't spend a lot of time on stuff you already should know."*
Understanding Students	There are characteristics of the students who come into the corequisite course, but there is also some understanding of the students that comes as a faculty member works throughout the semester. This theme highlights some areas of understanding that are discovered as a faculty member works with their students over time. Examples: *"Maintaining the motivation of the students is difficult."* *"The corequisites course allows the ability to assess students' understanding."*

Themes were developed from interviews with faculty who teach or have taught the corequisite mathematics classes. The list includes the description of each theme.

To further explain the results of the interviews, the theses are discussed in detail below.

Blending Courses

A corequisite course is a course taken concurrently with another course. Therefore, students are enrolled in two courses in the same semester. The theme that emerged in analyzing the interviews centered around the issues of trying to teach blended courses. This theme is centered around the comments shared concerning the blending of these two courses. A reoccurring comment was the concern of having the two courses blend into one. It is a "hard balance" because of the desire to maintain "class consistency." Whether it was one instructor or two, in both situations these instructors expressed concerns over their ability to balance the instruction in both courses.

Instructors wanted their students to feel the blend of the two courses and feel like they were ultimately enrolled in only college algebra, with support and some fundamental aid when needed on certain algebraic topics. For the faculty, it felt like a "consistent shift from developmental material versus the college level material." The most difficult strategy that was mentioned by the instructors was having to "teach so many concepts within the scope of both courses." The courses taught by two instructors might be presented using many different methods, which can be confusing for the students if the instructors are not aligned with how they approach the teaching of the class. This challenge did not always cause a negative situation.

In College C, where both teachers taught in one class, blended courses "offered a way for students to see the connection between two different methodologies" and provide them the opportunities to choose the more effective methodology for them. Another positive view on co-teaching mentioned was "the ability to have multiple events happen in class." A class with two professors gave the students an opportunity to "receive instruction and activities at the same time." One professor could "teach, while another graded a quiz, providing students with opportunities to receive instant feedback on assessments." One professor "could teach, while the other offered individual help." Having two professors provides students with "constant feedback on assessments for the class," because there is "always an instructor to observe the students for understanding." Blending the course overall is a challenge, but it can be balanced with the modality of how the class is taught and how the instructors are utilized.

Characteristics of Students

Students in a corequisite course have a wide range of characteristics based on feelings, skill levels, self-esteem, and background. Within this theme, the characteristics of the students observed will be discussed by instructors in corequisite courses. At each college, the feelings of the students upon entering the class were described similarly. "Low confidence in math" was an echoed phrase among the college instructors interviewed. Community college students can be unique because many of them come from a "complicated life," as stated by one interviewee.

Many students are returning to college and are far removed from any type of mathematics course, so they may lack fundamental skills. The combination of life and classes leaves students in a situation where they are "just overwhelmed by everything presented in class." Another instructor agreed with the idea that students struggle with "how to be a student." One instructor stated that "students lack time management, note-taking skills in math, and to add even more barriers, they are taking too many classes." The corequisite course requirement puts students in two mathematics classes, but students are also taking other non-math classes, which may not be the best idea for students. A student may think, "I will take a math class, an English, and an elective" and that sounds reasonable; however, the math class is TWO math classes. The combined load of the two math classes is equivalent to the load of two separate courses.

This concept may be misunderstood by students, which will be discussed later in another theme. There were positive comments from the instructors related to this theme. Some instructors stated that, even with all the concerning characteristics of students, they still enjoyed teaching in the corequisite model because "it is a better situation for students because they seem to be all at the same level math-wise." The less variety you have with regard to levels in a class, "the easier it is for an instructor to pace and teach the course." Additionally, one instructor stated, "If the students are dedicated to the course, they will do well." It seems in corequisite courses, there are students on both extremes, according to the instructors: "the dedicated ones and the ones that are too overwhelmed to dedicate what is needed in the course."

One instructor stated, "If a student is dedicated, regardless of math ability, they can find success in this type of course." More students in corequisite courses seem to "have an ability to be dedicated, even if they lack the basic fundamentals of mathematics." Thus, dedication is possibly needed more than skills.

Corequisite Course Characteristics

Corequisite courses possess a collection of characteristics that shed light on their unique environment. Through the discussion of this theme, a list of some characteristics professors feel make the corequisite courses unique is presented. Within this theme, instructors discussed both the positives and negatives of corequisites. Some positive views discussed were that the corequisite course in their college provided an environment to build on knowledge because "you are always reviewing and having a place to build the confidence in mathematics"; it is an opportunity to "deep dive into the developmental mathematics that is needed for college algebra."

Table 3.4.

Blending Courses		
Positive	Negative	Neutral
Many Different Methods	Hard Balance	Alternating Workload
Multitask in One Class	Students Favor Specific Method	Consistency Check
Multiple Events in One Class	Course Alignment	Want Blended Experience
	Lack Seamless Blend	
	Two Class Characteristics	
	Two Separate Classes	
	Grading Not Aligned	
	Class Not Consistent	

A corequisite course is a course taken concurrently with another course. Therefore, a student will be enrolled in two courses in the same semester. Table 3.4 categorizes comments shared concerning the blending of these two courses.

It was discussed earlier that students lack math confidence; one professor from College C mentioned that the corequisite course "is basically a place where students can take time to build that confidence." One of the other themes not being highlighted was the *standalone remarks*; one thought discussed was the lack of ability to "build on" with the students in the standalone classes. A standalone college algebra class may not contain the components for building confidence; instead, most of these need to be completed outside of class by reviewing material for students to study on their own time. One characteristic that can be seen as a great bonus for students is that during corequisite classes students "can have that safe space and time to develop as a student." Another benefit mentioned was the ability to "implement various techniques."

As stated before, various techniques can be a hinderance to someone's learning, while using other pedagogical strategies can be a tremendous bonus "if the time is available to dive into the multiple ways of solving algebra problems." Through the corequisite environment, there is "time to develop a fun, supportive community among classmates," which College C took full advantage of within their co-teaching model.

Some of the expected negative aspects of a corequisite course that were mentioned by the interviewees were the intense workloads. Unfortunately, "it is two courses, and it could feel like two courses worth of work, especially if you have two different instructors." It is "a lot of work," as several instructors mentioned. One instructor flat out said, "it's just too much." Along with the work, the classes, especially "preparation for college algebra, requires a heavy amount of complex computations within the problems." Even though the practice should help boost confidence, "there is no doubt it is an extreme amount of work to keep up with both sides of the math courses in one semester." Overall, in summary, the more in depth one looks into the list of aspects connected with corequisites, the more positive the characteristics are, while the negative side contains areas that can be addressed to improve corequisite courses.

Course Misconceptions

When students are placed in a corequisite course, they may not fully understand what they are signing up for. Within this theme, instructors shared some thoughts that confirm many misconceptions students have about corequisite courses. The first day of school is "always filled with questions about the course," and for a corequisite course, the questions are usually double that amount. During the interview, instructors from every college mentioned the repeated "misunderstanding of the course and also misunderstanding the time commitment."

The students are taking two math courses, but will they understand that requires two sets of homework problems? Will they understand it takes time outside of class for both classes? From interview results, it is common that this understanding is not grasped by those who register for the two math courses. "The realization of taking two classes sinks in as the semester proceeds, but it's definitely not understood during the first weeks," an instructor from College B stated. The time commitment of the courses is the biggest takeaway from this theme.

However, there is a silver lining according to professors as well. Some corequisite courses have "some unexpected fun and support." Some students were not prepared to "have a mathematics class that has as much support as a corequisite one does." Because of the time the group spends together, there is a "sense of community that is formed in this type of class." There were three instructors who mentioned this as being a "phenomenal advantage" of the corequisite model. The professors find value in the corequisite for students as they "gain that respect for mathematics."

Table 3.5.

	Characteristics of Students	
Positive	Negative	Neutral
Retain the Concepts	Additional Non-Math Courses	Two Types of Students
Student Levels the Same	Years Removed from College	
All Students Tested	Missing Fundamental Skills	
	Lack Reading/Analyzing Skills	
	Complicated Life for Students	
	Took Course Multiple Times	
	Multiple Attempts at Corequisite	
	Missing Basic Arithmetic	
	Low Confidence in Math	
	Not Utilizing Resources	
	Difficult Load of Courses	
	Delayed Response	
	Not College-Ready	
	Need Developmental	
	Need Fundamental	
	Far Removed from Math	
	Student Overwhelmed	
	Returned after Break from College	

Students in a corequisite course come into the course with a variety of characteristics based on feelings, skill levels, self-esteem, and background. Table 3.5 addresses some of the characteristics common to students in corequisite courses.

Faculty Expectations

A faculty member who teaches a corequisite course, especially for the first time, may have a set of expectations for the course ranging from classroom-related to student-related expectations. Whenever there is a new mandate with schools, educators have the sense of "learning as you go" and "tweaking the class as you go," as stated by two instructors from the interviews. Some instructors want the corequisite to "feel like a purely standalone college algebra course with fundamental aid when needed." Because of the amount of time spent in the courses, some instructors want "the group to feel like a cohort as they push through the course and build relationships."

On a downside, there are moments where instructors stated they "feel an extreme amount of pressure to get the students through all the material in one semester." The amount of content to cover in the two classes can be overwhelming, not only to the students but to the faculty as well. However, an interesting quote mentioned in the interviews was the mantra saying, "quality over quantity learning," because there is "so much to cover." It is "better to have students understand less material fully" instead of completing all of the material while the "students do not have a grasp on the lesson to make sense of the next course level." How instructors use resources can actually predict how faculty feel about the course.

One of the other themes, *corequisite resources*, included statements from instructors such as "no guidance" and "lack of preparation time." These feelings can definitely contribute to how instructors feel about the expectations of the course. Another mentioned "book integration" and how much it has "helped with constructing the two courses." The layout of the book helps with the layout of the course and what to expect as deficiencies in a college algebra lesson. Book integration is helpful because, as one instructor mentioned before, "most students are in the same level as far as their skill level" (meaning, there is less variation in skill level). Overall, the idea of "adjust as you go" was discussed by the instructors with regard to their expectations for teaching a corequisite course.

Table 3.6.

Corequisite Course Characteristics		
Positive	Negative	Neutral
Build Knowledge	Intense Workload	Test Review
Slower Pace	Lots of Work	
Build Confidence	Not Easy	
Just In Time	Heavy Computation	
Confidence Boost		
Various Types of Reviews		
Students Do Well		
Not Overwhelming		
Build Every Day		
Instant Feedback		
Little Bit Every Day		
Frequent Testing		
Always Reviewing		
Deeper Dive into Developmental Math		
Always New Stuff		
Mandatory Tutoring		
Individual Student Attention		
Various Techniques Shown		
Refresher Good		
Help Build Student		
Fun Community		
Deeper Relationship with Corequisite		
Building Supportive Community		

The corequisite course has a collection of characteristics that sheds light on its unique environment. Table 3.6 lists of some characteristics that make the course unique.

Format of Course

The format of how to teach a corequisite course is not consistent across the United States. Many schools, especially in Texas, have been using a trial-and-error method as they try to create the best environment for these students. Instructors addressed some of the various formats during interviews. When it comes to the format of the course, there are elements of the course that seem to have an impact on the success of students, either in a positive or negative way. College C's, two professors co-teaching both the corequisite and the college algebra course, discussed the format that they "generated that has the most praise from faculty."

Having two teachers in one class has far outweighed the other formats in positive comments throughout the interviews. This format "allows flexibility in the classroom," "room for deeper interaction," "quick assessment," "instant feedback," and more "thoughtful thinking about how to teach students." The professors on every campus shared the idea of teaching in a "Just in Time" format, meaning addressing developmental lessons right before they are needed in college algebra. It seems to be the most effective. One professor at College B mentioned that "once the developmental lesson is taught, it is informally or formally assessed throughout the semester." This technique helps "the students not lose their understanding of the material."

However, one mentioned "the downfall for co-teaching is not finding the right partner" when co-teaching with another in the same room. There were a few professors who found instructional collaboration to be a negative experience due to "not having enough time to collaborate." Preparation time with co-teaching is a very important necessity to ensure that professors are on the "same page with goals, expectations, activities and assessments." Overall, a team of instructors can have a positive impact on a corequisite course if planning time allows for the establishment of a collegial atmosphere where both individuals have time to plan effective lessons for students.

Online courses have been an option for corequisite courses, and the feedback shared with the instructors who teach the online version of corequisites is that the students benefit more from a "synchronous online setup." A synchronous online course requires the entire class to be online; however, the class has virtual meeting times similar to face-to-face classes. All the instructors who taught the online courses "preferred this format," if teaching online, to "keep students engaged." With an asynchronous type of class, there is no meet-up time. One professor at College A mentioned how "delayed a response will be from a student in an asynchronous course because they get lost in the virtual with little live interaction."

Mixing modalities has been another difficult experience among professors. There are some colleges that still have many virtual courses, but the colleges are trying to shift back to more on-campus classes. At College A, corequisite courses are "split in half to let some students come to class face to face for one day a week and then watch the class virtually for another day of the week." As an example, a student may come to class Monday, but on Wednesday they participate in the class online. This technique is used to maintain social distancing in the classrooms. However, this method has "caused stress for faculty" according to the interviews and is "not preferred." In some places, format does matter in how the professors feel about their job, which leads to the next theme.

Table 3.7.

	Course Misconceptions	
Positive	Negative	Neutral
Unexpected Fun	Misunderstood Course Description	
Fun Exercises	Misunderstood Time Commitment	
Unexpected Understanding of Math	Expectation Not Clear	
	Realization of Two-Course Load	
	Commitment to Both Courses	
	Miscommunication to Students	
	Students Don't Understand Necessary Commitment	

When students are placed in a corequisite course, they may not fully understand what they are signing up for. Some misconceptions students may have about the course are listed in table 3.7.

Job Satisfaction

When teaching a corequisite course, a faculty member may often wonder how the change in the developmental math courses has affected job satisfaction, and even simply how faculty feel teaching within this new course model. This section will present the feelings of faculty as they reflect on corequisites and their work environment. Every single instructor expressed confidence in the direction that corequisite courses have taken and some even stated that there are "many advantages to the new course model."

Most instructors would even go as far as saying they "prefer corequisite classes" over the standalone courses. Regarding the format of the courses, co-teaching has produced some of the most positive feedback when it comes to job satisfaction. "I feel like I am on a winning team and the corequisites were designed in the right way," says one instructor at College C. Two instructors at College B mentioned how "positive and flexible the corequisites can be with the additional time granted." Instructors discussed how "corequisites offer an abundance of job satisfaction where work is now fun." These positive feelings are "transferred to the students," which is important. The "students sense everything," said one instructor at College C.

They know when "an instructor is not happy with their job." Being able to find the "peace in the change is the beauty of the situation." These positive comments came from a sense of understanding that there have been significant changes with developmental education, but these instructors have found a way to embrace it. One instructor stated how this change has made them feel "they were confident in their ability as a professor."

Table 3.8.

Faculty Expectations		
Positive	Negative	Neutral
Textbook Integration	Preparation Time for Teachers	Do the Best You Can
	No Guidance	Expected Prerequisite
		Set Expectations
		No Negativity in the Class
		Tweak Class as You Go
		Interactive Assessment
		Cohort Feel Among Students
		Algebra with Support
		Teachers Learn as They Go
		Helping Student Success
		Expected to Get Them Through
		Outside Resources to Help
		Quality Over Quantity Learning
		Student Learning Outcomes (SLOs) Provided
		Computer Program Assistance

Table 3.8 shows some of the expectations of faculty when it comes to teaching corequisite courses.

Time

Corequisite courses have a connection with time that was discussed throughout the interviews. The "extra time has been the single [most] beneficial part of this mandate of corequisites." Even though some instructors feel like there is "too much material to cover," everyone overall finds a "benefit of having extra time to share with students." It allows for time to "build a rapport" with the students.

The students who are "not successful are those that do not dedicate the time that is needed, especially those that have attendance issues." When students "take advantage of the extra time that is granted when signing up for a corequisite, they find success in completing the courses." The combined format of the course means that the "more days a student spends a week on math," according to the instructors from College A and C, "is more beneficial." Classes that meet 4–5 days a week are stronger than those that meet two days a week because the professors feel "more days, more beneficial" as stated by an instructor from College C in the interview. Time together is valuable.

Table 3.9.

Format of Course		
Positive	Negative	Neutral
Just In Time Teaching	Teaching Different Modality	One Teacher Both Classes
Two Teachers Add More Attention	Limited with One Teacher in Class	Course Depends on Format
Two Teachers Helpful	Different Teaching Formats	Two Teachers One Class
	Lack Connection with Co-Instructor	Not Online Only Course Format
	Challenge with Asynchronous Classes	Paper/Pencil Assignments
	Instructor Collaboration	
	Working with Another Instructor	
	Concern with Corequisite Students on Next Level	
	Not Prepared for Next Level	
	Class Pace Concerns	
	Class Pace Balance	

The format for how to teach a corequisite course in institutions of higher learning is inconsistent across the United States. Table 3.9 notes the pros and cons of various formats that were discovered in the interviews.

Understanding Students

There are characteristics of the students that are an integral part of corequisite courses, but there are also some understandings of the students that come as a faculty member works throughout the semester. Discussed here are some areas of understanding that faculty members discovered from working with their students over the length of the semester. Faculty members discussed expectations, and how some of those expectations were suppressed as they learned more about their students. Corequisite courses provide "opportunities to assess students' understanding" of mathematical topics and instructors learn how to "adjust to what these students know and do not know."

Faculty expectations are that students have "some sort of mathematics prerequisite skills," but they may discover later that students in their class "lack many prerequisites." Some instructors argue that the students in their courses "have a wide range of skill level[s]," but upon further conversation in their interviews, they admitted there are "much higher variations in the standalone college algebra classes." Placement is an overall issue and that is a problem that is addressed in the *administration issues* theme. However, administration issues do add to the faults and errors in placement tests. Unfortunately, many students are placed into a college algebra standalone class but lack many of the skills needed to succeed (as mentioned in chapter 1). Understanding students is "stronger than any placement test," especially when there are skills beyond mathematics that contribute to success in the course.

Understanding students may lack college preparation skills is extremely important. During the interviews, faculty discussed how they "made significant discoveries working with the students in corequisite courses," and the common code that emerged in the theme was how "students lack study skills and note-taking skills." One professor from College C stated, "some students that are far removed from college have forgotten how to be students." Additionally, even recent high school graduates are "not demonstrating the necessary skills to successfully pass a corequisite course." Students need "study skills activities that can assist them in making them better students."

Listening to the Voices of Faculty of Developmental Mathematics 43

There were some comments made about "dual students not being corequisites and corequisite students not being college-ready." This discovery was made by the professor from College B when discussing how a "student handles the workload." This discovery provides the instructor the knowledge to be proactive while evaluating the prior knowledge of his or her students early in the semester. These tools are life-changing for students starting their college careers.

Table 3.10.

	Job Satisfaction	
Positive	Negative	Neutral
Teaching Schedule Favored	Too Much	First Time Teaching Corequisite
Finally Got It Right	Exhausted	Purpose of Assisting the Struggling
Work Is Fun		Students Sense Everything
Magic in the Classroom		Teachers Should Want to Help
Prefer Corequisite Class		Teachers Make a Difference
Wish to Only Teach Corequisite		
Many Advantages		
Many Successes		
Strong Believer in Corequisite		
Praise for Corequisite Class		
Teacher Schedule Preference		
Learn Better Ways		
Corequisite Model Works		
On a Winning Team		
Love Corequisite		
Corequisite Designed in Good Way		
Teacher Never Stressed		
Confidence in Ability		
Enjoy Co-Teaching		
Positive and Flexible		

Table 3.10 is all about the feelings of faculty when it comes to teaching corequisite courses and in regard to changes in their work environment.

RESULTS

The combination of themes discovered brings some interesting perspectives to mind on how faculty feel about corequisite mathematics and teaching. Tables 3.4–3.12 are organized in a way to separate a positive effect and a negative effect of particular codes under each theme. There is also a neutral column that contains some codes that appear to have no general effect. While investigating those codes, some thoughts can be mentioned; however, the study is careful not to generalize to the whole population of corequisite faculty members with a small sample size.

One can view these codes and themes to relate and compare how faculty generally feel about teaching with a corequisite model. One can observe some similarities or even discover similarities that they may not have noticed until viewing the tables. One instructor from the interview had a moment where they "never thought about that until you mentioned it, but that is a great point." Realization of some positives, or even some negatives, can come to the surface after sharing these types of experiences.

Generally, there were more moments of positivity when it comes to developmental and corequisite education. Even though in some parts of the country educators are promoting significant changes in developmental mathematics, to the pressure felt by faculty as a result of these changes can be added a silver lining of hope in making this a positive change.

From the data collected, it appears that the corequisite model overall is a good one, even with the "intense workload" and "heavy computations." Instructors of corequisite courses voiced mostly positive comments from "building confidence with students" to "just in time teaching." The corequisite courses offer "more time to practice more" and "time to connect." Even though some professors committed to "too many class meetings," the benefit of those class meetings seems to "build a better rapport with students." It was stated many times that "more times is more beneficial," even though students may struggle with the idea of longer classes. The time is valuable in building students' mathematical confidence and understanding of concepts taught in these mathematics courses. Many students have a combination of characteristics that can hinder their success in their corequisite mathematics classes, and those codes were reviewed as well. However, these characteristics are from the perspectives of the faculty members, and a study examining student perceptions would have to be conducted to substantiate these claims.

One idea voiced by professors heavily suggests that students have many components that stand in the way of their success in the classes. The most common complaint is students not being college-ready in the sense of "knowing how to be a college student" and this realization has nothing to do with mathematics skills. Students seem to struggle with understanding how to "utilize resources" provided in the classes and with "note-taking skills." Note-taking in a mathematics course can be an overall struggle for many students, because more writing is done in working out problems and not necessarily in taking notes.

One instructor noted, "Students are not being taught how to read a mathematics textbook" and it is not surprising that taking notes would be such a struggle for these students. Some students are "years removed from college" or any classroom setting. Therefore, taking mathematical skills out of the equation, students seem to struggle with time management, study management, and organizational skills that help with the success of a college student. To add to the pressures of being a college student, there are the pressures from outside life; these times are presenting "complicated lives for students." A complicated life in addition to a "difficult load of classes" does not equal a thriving situation for students. The positive side is that if they dedicate the time to the course, students can be successful.

The skill level does not matter; in fact, the popular response from the instructor interviewees is that the skill level is widely consistent in the corequisite courses. "You know what you are getting and can adapt to that easier," said an instructor at College A. They are "all testing and student levels don't vary as much as my standalone classes," said an instructor from College C. Placement tests have often been criticized for not placing the students where they need to be academic-wise but for corequisites, it seems to be a great starting point for most students placed in the course.

As the instructors teach in the "just in time" fashion, the fundamentals needed for each section seem to be needed by most students in the corequisite class. To add to this positive perspective, there are the additional times available to review those fundamental concepts in corequisite courses.

The format of the course played a major role in how faculty felt about the characteristics of the corequisite course and added to their overall job satisfaction. College C has a co-teaching model set up. This meant both instructors were present for both parts of a corequisite model and shared responsibilities in class. "We finally got it right," was the comment of one instructor as they raved about how "work was fun" as they created "magic in the classroom." The format of the course provided instructors opportunities to learn different techniques and "learn better ways" to make the corequisite classes work.

College B is experimenting with two instructors, but they are teaching their own courses—one teaches the corequisite and the other teaches the college algebra. This format sounds like a great idea but having the teachers separate leaves a disconnect between the teachers if they are not taking the time to collaborate for the courses. If the teachers were forced to connect more often, it might create a better experience for the instructors. "There was not much time to collaborate," said one instructor. Co-teaching where both instructors are present in both classes creates a situation where the instructors are required to interact, and they interact often.

College A has single instructors, and their format was favored as they stated, "I almost prefer corequisite courses," but the work seems overwhelming for both instructor and the students as they navigate through the course on their own. "Tweak as you go" was a remark made by one instructor. One format that is difficult is the idea of having a virtual and face-to-face class combined. Handling face-to-face students and online students adds a level of stress for instructors; this format is popular in light of the pandemic and keeping social distance standards in the classroom.

Format of the class plays a major role in job satisfaction, because when the format is right, the instructor feels positive about their job on a day-to-day basis. When the format is separate, the comments from the interviews are hard to hear. It is "just too much" and I am "exhausted." The class is viewed as a "hard balance" as instructors try to understand what the students need and how to assist them throughout the semester. The format matters and seems most likely to contribute greatly to professors' level of comfort with teaching corequisites.

Another aspect to investigate is the comments of the math instructors versus the math education instructors. For clarity, math instructors have a purely mathematics background without a mathematics education degree or a teaching certificate from teaching at a K–12 level. Math education instructors either possess a mathematics education degree or have obtained a teaching certificate from teaching at the K–12 level. The study sought to interview one of each instructor type at each campus to determine if there were any overlapping patterns or themes in perspectives.

The results suggest that most comments relating to a positive feeling of job satisfaction came from the math education instructors. With these faculty members interviewed and regardless of campus, there is a sense of favor and love for corequisites among math education professors. Most comments about the difficulties of the format and the blending of the courses came from math instructors without an education background. Their negative thoughts were mentioned many times. This suggests that, for these math instructors, structure and alignment is important and the feel of "having two different classes" is not viewed as a benefit. Math education instructors, on the other hand, viewed the extra class as "extra time to build."

An interesting scope after reviewing data showed that, even though the math education instructors had a more positive stance on the corequisite courses, the pure math instructors had more positive understanding of their students. Discovered during the semester with the students, the data shows that the math instructors gained deep understanding of their students while teaching the course.

The ability to truly understand students is a valuable skill to develop. Even though there are characteristics that students of corequisites possess, there are also many characteristics that are only discovered while spending dedicated class time with them. This is not to suggest that every college instructor's experience will be the same, but it does offer some interesting perspectives to think about with the group in this particular study.

QUALITATIVE TEAM CONSISTENCY

The interviews were analyzed and coded by a team of qualitative researchers, two PhD students, and a doctoral recipient by way of qualitative research. The group explored the data individually before coming together to combine the results. From there, the codes were grouped and determined the themes that had emerged upon frequent reviews of how to associate codes with each other. The main themes credited were then shared with the group.

Discussion took place of how the themes were associated and how they differed, how the definition could be polished and clarified, changes that could be made to the table of definition to clarify themes for readers, and any additional changes that might be needed. Once this list was solidified, a sample interview was used to test the themes created. One interview was split into lines and distributed to each person in the qualitative group. Each person was to pick a theme for each line in the interview. The overall goal was to see how many lines that we were consistent at picking had the same theme.

The results were that out of 103 lines, 80 lines had our group pick the same theme while 23 did not. This result gave us a 78% accuracy on matching the themes and understanding how the words paired with the theme. Even though, with the coding, we were not trying to quantify our results, we did want to make sure that our thinking was on the same path as to how we felt about the words from the interviews. The consistency check added some sense of value and validity with the themes we picked and how they correspond to the words of the participants.

SYLLABUS REVIEW

Upon reviewing the syllabi of the interviewees, there are some key elements that were necessary to point out about the format of the course. There were several consistencies with the syllabi, like the credit amount and student learning outcomes. These items are standard with the corequisite model due to Texas state mandate. Not surprisingly, there were some differences in grading that were of some significance in sharing. One college had a heavy weight on exams and a large amount of exams; 85% of the grade was based on exams and 15% was based on homework.

Listening to the Voices of Faculty of Developmental Mathematics 49

There were seven exams for the college component. The intriguing option was that if a student failed exams 1–5, they had the opportunity to discontinue with the college algebra portion of the course and just focus on the developmental portion. This characteristic was interesting because this college also had a heavy number of codes for the themes *understanding students* and *time*. The idea of making this an option for the students indicates an understanding that a student has a chance of passing part of the two courses if they focus on one course.

Another college put a heavy weight on homework and in-class assignments for the developmental portion of the corequisite class, but the exams weighted heavily in the college algebra portion of the corequisite class. This idea shows the importance of practice with the material and helps with blending the courses. Reviewing the syllabi of the interviewees reveals a large amount of consistency but offers variety in grading in a way to benefit the students.

Table 3.11.

	Time	
Positive	Negative	Neutral
More Time	Too Many Class Meetings	
More Time Together	Attendance Issues on Both Sides	
Devoted Time Equals Success	Neglected Time Equals Unsuccessful	
Dedicated to Class Equal Success	Not Enough Time	
Devoted Time		
More Days More Beneficial		
More Class More Practice		
Time Together		
Build Repair Because of Time		
Time to Connect		
More Days Equals Better		

Table 3.11 shows some of the comments made about time when it comes to corequisite courses.

Table 3.12.

Understanding Students		
Positive	Negative	Neutral
Assess Student Understanding	Maintaining Motivation Hard	Feedback Is Important
Test Format Matters	Lack Prerequisites	Learning Curve for Students
Address Different Levels	Wide Range of Skill Levels	Can't Treat Like College-Level
Attention Equals Success	Students Not Acclimated to College	Work on How to Feed Topics In
	Lack College-Student Skills	More Developmental Math Needed
	Absence Equals Failure	New Concepts to Students
	Need Prerequisite Before College Algebra	Dual Students Completely Different
	Need to Learn to Be a Student	
	Build Comfort	
	High School Lack Preparation for Students	
	Lack Note-Taking Skills	
	Student Development Is a Problem	

Table 3.12 highlights some areas of understanding that are discovered as a faculty member works with their students over time in a corequisite course.

PROFESSIONAL DEVELOPMENT

The question asking about whether campuses were equipped with professional developments geared toward helping with implementing corequisite mathematics courses was unanimously answered with a no, not specifically for corequisite mathematics courses. The purpose of this book is to understand some of the challenges that faculty of corequisite courses face and how to address these challenges with professional development that, from the interviews, currently seems to be a missing ingredient in the effort to improve student success in developmental education.

At College B, a respondent mentioned some professional development that is almost "peer-led, where [a] professor can share their experience of teaching the course, but no research-based or trained facilitator of a professional development" was assigned to instructors of corequisite courses. The interesting epiphany is that some of these campuses do offer some type of professional development for math instructors of corequisites, but the instructors were not aware of it. There are resources available, but the marketing to make sure instructors are aware of them may not be the best. For example, Texas has statewide professional development resources available, but it appears that the reach of those resources is missing some campuses. In chapter 4, we will explore some research-based professional development that can address some of the challenges found in the codes and themes of the interviews conducted.

Chapter Four

Successful Professional Development Options for Community Colleges

> Logistical challenges include . . . developing meaningful faculty professional development and achieving faculty buy-in.
>
> —Emblom-Callahan et al., 2019, p. 4

WHAT CAN WE DO?

What is professional development? It is the opportunity for personal growth. It is the "continued training and education of an individual in regard to his or her career" (Campos, 2020, para. 2). Professional development is an opportunity for faculty to improve in many areas, such as content knowledge, pedagogy, and dispositions (Burrows, 2015). Exploring various forms of pedagogy can help faculty members create additional materials and develop new approaches to use in classroom teaching (Burrows, 2015). Research shows that pedagogical learning is more effective for students, so training that helps faculty with improving this style can increase student success rates.

How this training is conducted is very important to making sure that faculty receive the best enhancements for their teaching. With all colleges and universities set at over 75% corequisite model with their developmental courses, a push for more professional development would add value to programs on individual campuses. As explored in chapter 3, the nine themes created from data analysis were the following: (1) characteristics of students, (2) corequisite course characteristics,

(3) faculty expectations, (4) format of course, (5) job satisfaction, (6) course misconceptions, (7) time, (8) blending courses, and (9) understanding students. Corequisite courses share some common issues for classes and faculty to address, but can the training be designed in a way to help with these challenges?

This chapter will attempt to provide some tips and suggestions for a campus that experiences some of the same trials revealed by our interviewees. Though we are not trying to overgeneralize any aspect of this study, we offer options to help campuses with backgrounds similar to the anonymous colleges shared in this book. As we explore some tactics that have helped other campuses, one can match these programs up with the themes found in the interviews.

CHARACTERISTICS OF STUDENTS

After reviewing the characteristics of the students in the interviewees' classes, there seem to be some common thoughts about areas where our students may fall short; however, it may be no fault of their own. It may simply be a lack of preparation for these students to enter college on a solid foundation. "One of the pressing issues facing secondary and postsecondary mathematics educators is students' difficulty making the transition from high school mathematics courses to college courses" (Frost et al., 2009, p. 227).

Another pressing issue facing students is the amount of time a student may have been away from the classroom. All of these things hinder a student's success in their classes, and it can have an enormous impact on a student who has the pressure of two mathematics classes at the same time. As an initiative in Washington, a group of researchers decided to do a program where they brought together mathematics educators from various platforms to discuss ideas to help their students. This program brought together teachers from universities, community colleges, and school districts to share and develop ideas that would better prepare students for the transition into college.

The group also was equipped with a facilitator to help drive conversations and handle conflicts that might arise. The group prided itself on leaving negativity at the door. In that particular area, there seemed to be a minor blame game between high school teachers and college faculty—the high schools blamed colleges for using outdated techniques and colleges blamed high schools for not teaching content-heavy math (Frost et al., 2009). This situation may be felt across many areas as we all try to find a reason for why students are not ready for college and may not be successful in mathematics overall. There are two sides and there are pros and cons taken from both sides; the goal of a program like this one was to find a solution rather than to lay blame.

The program involved forming a professional learning community (PLC) to examine school curricula and student work, to prompt discussion, and to share research. Meeting and reflecting are a major part of this collaboration. This helps teachers gain new perspectives on teaching and learning mathematics and new perspectives on what students could develop to facilitate their success in college. The main source of resources for the PLC was the College Readiness Standard, which included standards on math content, processes, and student attributes.

Over time, the PLC created change in how mathematics was taught in high school and college; and the collaboration also put students in a better position for mathematics in college. However, one should understand that this program was not an instant buy-in. Even the educators who took part in the PLC and developed new techniques were not immediately driven to change. Change is "likely to require that teachers examine and reconsider long-held beliefs about mathematics teaching and learning, a process that is typically slow to occur and difficult to accomplish" (Frost et al., 2009, p. 233). The change takes time and patience, and the result is a better understanding of students' experiences in both school settings. PLC is a great way to combine efforts and try on a different lens in education. Not only are educators learning, they are also sharing, thereby building a strong community for mathematics for our students.

COREQUISITE COURSE CHARACTERISTICS

Another characteristic of having a student ready for college is developing their note-taking skills, which was a repeated concern of the interviewees. Note that "many students may not develop good note-taking skills unless they engage in directed concentrated practice" (Eades &

Moore, 2007, p. 19). Good note-taking promotes active learning and listening in the classroom, and also increases motivation to want to learn. It can ease math anxiety and frustration, which is a characteristic of some math students (Eades & Moore, 2007).

An unexpected benefit is that it regulates the instructional speed. How many times have students said, "The instructor goes so fast"? Note-taking can help regulate that speed and also prompt the student to ask questions for clarity. This puts the student in control of the pace. Also, a good note-taker has their own review sheet at hand. Lastly, with a set of good notes, the student has resources that they can present to a tutor. When a student visits a tutor, a frustrating thing they may encounter is how the tutor may show them a technique that is different than their instructor's. If a student has quality notes, the tutor will have a guide to what the student learned and the tutor can help in a productive way (Eades & Moore, 2007). Showing mathematics students how to take notes is a skill that educators should share with their students as they start their college journey.

Note-taking strategies are presented in six stages shared in a book by Robert Gerver, who gathers some tips on how to enhance a student's ability to take good notes in a mathematics class. These strategies would be helpful for a professor to use in their class to teach students about taking good notes. The six stages of note-taking are the following: (1) No writing at all, (2) no annotations, (3) in-class annotations, (4) at-home annotations, (5) balloon-help annotations, and (6) math author project (Gerver, 2018).

Not writing notes is a common stage with students, especially if it is not mandatory to take notes. It is not the wisest choice, because when you take notes, you are "mentally processing the information" when you are able to write down something in class (Gerver, 2018, p. 9). Not taking notes puts students at a disadvantage. The next stage is "no annotations"; with this stage, the student will copy the notes word for word just as the instructor does in class. This method is a good first step, but not always the most effective in retaining information about the lesson.

The biggest issue with doing this stage only is that most students are not able to explain these notes if they are presented to them weeks later. There is not personal touch to the notes, but it at least documents the lesson. The next stage is in-class annotations, which is one of the most impactful adjustments to note-taking. This stage is where you add words from your teacher or other students that add a voice to the notes. It also helps with regulating the class pace, because it involves you asking questions or asking others to repeat certain parts. It keeps you active with the notes and is not just a mechanical-like copying of notes. This stage is done actively in class.

To add even more value to the notes, the next stage of at-home annotations is completed at home as you review your notes. This stage should be done on the same day so it is fresh in your brain and you can add any additional segments that you may not have included during class. During this stage, the student is able to pull notes from books or other resources to add to the notes taken in class. It adds connection to the student's understanding. When the student reaches a point of uncertainty, they can utilize the balloon-help annotation. With this stage, the student would add balloon annotation that can visually stand out in the notes.

Gerver suggests using animated cutouts to make these questions stand out. Balloon-help annotation can also highlight caution areas where mistakes are commonly made in the lesson. Lastly, and certainly not necessary, is the ability of the student to recreate the notes and exchange them with another student. The point is to see the value of the notes; the student is writing notes for someone who was not present in the lesson and seeing if they are able to follow. It shows if the student has gained the skills of true note-taking (Gerver, 2018). Again, this last stage is not necessary, but it would be useful to do for a few sections to gauge the depth of a student's note-taking skills. The stages are a great way to develop a blueprint of how students can create useful notes for their mathematics classes.

Another experiment shown in Gerver's book is comparing mathematics versus literature novels. He had two students compare two books: one was a mathematics textbook and one was a literature book. They both opened the book to a page, any page, and he told the students to stand far away (Gerver, 2018). The interesting thing was visually how different the two books looked. The math book was inconsistent with symbols and words not aligned; it was spotty. The literature book was smooth and consistent with the words and paragraphs.

The point was to show how the student's notes may look when one is dealing with taking math notes. There will be gaps and it will look inconsistent, but it has its unique looks, just as the textbook does. The demonstration showed how the textbook almost aligns with what your notes should look like. Reading a mathematics textbook may not be the most popular thing to do and that may add to the struggle to make good math notes. We lack clarity of what math notes may look like. Professors emphasizing the need for math notes can help build a student's confidence.

FORMAT OF COURSE AND BLENDING COURSES

To visualize how a developmental mathematics lesson can help with developing ideas and improving teachers' instruction to their students, a video can be a valuable tool in enhancing teachers' instruction; however, it does take levels of training and preparation to make sure the training is received in a meaningful way. The guidance will help the teachers not fall into their "existing conception of effective instruction" when watching the videos, which is common (Borko et al., 2014, p. 260). Facilitation is a very critical factor when it comes to using videos. In this article, the writers chose to focus on two different projects: the learning and teaching geometry (LTG) project and the project-solving cycle (PSC) project.

The two projects were completed using different approaches to show the readers the range of how video-based professional development (PD) can be used. The LTG is a highly specific project, where facilitation materials are distributed for a specific PD experience (Borko et al., 2014). The PSC is a highly adaptive project, where the goals and resources are developed from the generally guided facilitation of the group (Borko et al., 2014).

When thinking about the practices of video-based PD, the article discusses how the techniques reflect many practices in the classroom. Anticipating student responses is a result of practice and preparation. The way a teacher anticipates questions and directs the classroom involves the same approach that is needed in the PD workshops. Anticipating responses from teachers necessitates familiarity with the material. Monitoring the thinking of the students necessitates the ability to ask leading questions. Creating leading questions is a powerful skill.

To be able to develop that in the PD workshop, just as in the classroom, one will need "consistent, ongoing professional development and opportunity for reflection" (DeJarnette & Hord, 2020, p. 1590). With the monitored thinking, a teacher can navigate various approaches for the class to explore. Professional development, especially if video based, involved a selected approach—the selection of videos to show and the selection of conversations to mold the framework of the workshop. Sequencing students' shared work is another practice which echoes the needs in the PD environment when involving teachers. Lastly, connecting student responses to one another and to key math ideas brings the purpose of the classroom together (Borko et al., 2014). The purpose of letting the students explore, while tying it back to the objective and curriculum of the course, is what creates a powerful classroom for the students and likewise a powerful PD workshop for the teachers.

The article goes into detail about how there is a moment of planning and a moment of orchestrating video-based discussions. A skilled facilitator will take this opportunity to make sure he or she is prepared by reviewing video clips for discussion but take some time in selecting parts that are important for the workshop. The next step would be to construct questions that go along with those selected video clips. Now, the orchestrating is ready to proceed.

In orchestrating, the facilitator wants to be able to pull the thinking out of the participants. It may be a tough crowd; it may be an active crowd. However, the job of the facilitator, no matter what the crowd may be, is to draw conversation and discussion out of the group. Once the conversation is moving, stretch the group to match evidence to their thinking; stretch the group to solidify their case with facts. Lastly, to wrap up the moment, connect the discussion to key mathematical and pedagogical terms and ideas (Borko et al., 2014). The planning that is involved with video-based PD is the same type of planning that may be necessary in the training for facilitating PD for secondary school mathematics teachers using student work.

CHARACTERISTICS OF STUDENTS AND UNDERSTANDING STUDENTS

Viewing student work brings focus to common mistakes and areas for improvement in instruction. To help with forming professional development for faculty in higher education, we will look into the professional development of K–12. PD directed specifically toward secondary school mathematics teachers can be rewarding and challenging when it comes to including student work. PD facilitators have the ability to "engage secondary school mathematics teachers with student work in ways that afford powerful and potentially transformative learning opportunities" (Silver & Suh, 2014, p. 283).

Viewing student work can help with how teachers assess mathematics in their classrooms. There are two very different approaches to assessing in the classroom: evaluative and interpretative (Silver & Suh, 2014). Evaluative assessment is in light of right or wrong, where the teacher is looking to see if the student got the answer correct. Interpretative assessment is in light of understanding the student's thinking, where one listens more to the student's reasoning. The goal is for more teachers to be more interpretative in their approach, to push to better understand their students' thinking rather than just knowing if they got the answer correct or not.

Where are the challenges in PD with secondary school mathematics teachers? There is some study in this article that cautions researchers about using student work for secondary teachers. The problem shifts to the level and amount of mathematics a secondary teacher has received as opposed to a primary teacher. The idea is that many secondary teachers have had more mathematics courses and may have a calculational orientation to teaching. The article stresses that instructors at this level may not question mathematics as much because of their success in it (Silver & Suh, 2014). That success leaves little room for questioning the validity of mathematics. When someone is not as mathematics based or does not go as deeply into mathematics, there is opportunity for interpretation and curiosity; it leaves room for more questions.

An interesting case study gave a survey to a group of teachers to rank their expectations of their students on a set of problems (Silver & Suh, 2014). The predictions were poorer for the secondary teachers who had a static view of students' abilities and felt that the order in which mathematics is taught is important and should remain fixed, whereas the primary teachers had more open-minded students, grasping concepts using less mathematical thinking but more of a general understanding of life. Another problem is that more secondary teachers seem to be stuck in teaching their one or two levels but know little about the mathematics before and after that level (Silver & Suh, 2014). The minds are more open for the teachers in the primary school levels.

These challenges are some to consider for further research in this group. The warning is not to stop the research but to take into consideration the need of being aware of the approach to the group. Valuable data can be collected from this group, so the research is needed. This awareness is part of enhancing the experience as a facilitator; it is part of self-improvement.

BLENDING COURSES AND JOB SATISFACTION

This strategy is tied into improving the feeling of pressure with teaching and collaborating with other educators, and it is found in an article that emphasizes the power of collaboration to improve learning environments and increase interest in the sciences. Andrea Burrows wrote an article called "Partnerships: A Systemic Study of Two Professional Developments with University Faculty and K–12 Teachers of Science, Technology, Engineering, and Mathematics (STEM)" about the important role K–12 teachers play in increasing the interest and success of STEM classes by the way they deliver instruction (Burrows, 2015).

In the study, a yearlong professional development was conducted, connecting 31 K–12 teachers with university faculty to create a partnership for 19 days throughout a year to show best practices and strategies for teaching STEM material. The goal was to enhance the students' experience with STEM and, also, to make teachers more comfortable with including modern sciences in their classrooms. The results were impressive with improvements across the board in feelings about STEM (Burrows, 2015).

Burrows states the problem and issues in this article to show the need for professional development. The development of the program was based on research and valuable to both parties.

Professional Development should also concentrate on general and specific content knowledge, student objectives, and common student misconceptions . . . and include time for instructional planning, discussion, and consideration of underlying principles of curriculum [which] may be more effective in supporting implementation. (Burrows, 2015, p. 29)

The reason for the program was to promote more STEM majors and even stronger STEM majors; the need of this program in Texas is the development of corequisite models; the value of professional development is to increase support for new reforms.

The methodology of the research is stated in the article. The research questions included the following:

1. What are K–12 teacher perceptions/uses regarding the PD content provided?
2. What are K–12 teacher perceptions regarding the PD partnerships, especially in relation to discussion?
3. What are faculty perceptions regarding the PD partnerships with K–12 teachers? (Burrows, 2015)

These questions specifically tap into the views of the participants of the program and are open-ended to get qualitative data. This questioning was not just a checkbox survey to see if the program was successful—the professional development team wanted to know what worked; the researchers really wanted to hear the feedback and see what specifically worked and what did not.

This program had a profound impact, according to the data. On the inclusion of astronomy content in the classroom, pretest showed that only 16% of teachers found this useful. Posttest, the increase was to 84%. For connections and collaborations, pretest showed only 26% of teachers found this useful, and after the program, 90% found collaboration and connections useful (Burrows, 2015). Saying the content is effective in STEM, the results were 18% pretest and 82% posttest. Regarding whether teachers thought professional development had the ability to build partnerships, 26% pretest believed this was true and 90% posttest believed this was true (Burrows, 2015).

There are similar results throughout this article which also goes into the interpretation of the effects of the program. The program overall left a lasting mark in making teachers more confident in delivering STEM content, enhancing both their perception of STEM content and their views on collaborations. The same can be transformed into a professional development program for corequisite instructors.

JOB SATISFACTION AND BLENDING COURSES

Techniques on how to blend and co-teach courses can include research groups on the best approach to certain lessons. After reviewing video-based PD facilitating and study work PD facilitating, these can be pulled together to improve instruction. In this article, there is a discussion on having a teaching research group (TRG) to help with this instructional improvement. There is a framework of how this TRG operates, and it could combine both video-based facilitating techniques and study work assessing.

The framework of the teaching research group involves the following: prepare the lesson on the Pythagorean theorem with the group for the teacher, give the lesson to their classes while recording to observe later, conduct a post-lesson interview with the teacher, come together with the group to share feedback about the video, make necessary revisions to the lesson, give the lesson again while recording, share feedback on the second video, give the lesson a final time, and, lastly, share overall thoughts of the cycles (Yang, 2014). The practices discovered in the video-based article can help with dissecting the video recordings from each lesson; it would help with how to analyze and make the best use of the videos in this project. After the lesson, study work can be observed to see how participants absorbed the lesson and if they indeed took away the objective of the lesson. A combination of both concepts to improve facilitation can be key components that push this TRG to the ultimate level of professional development.

COURSE MISCONCEPTIONS AND TIME

Many campuses have learning assistance centers available to students to get extra assistance with courses. One study was done on these centers to understand how impactful they are on various campuses. The study took a collection of campuses and had them investigate their own centers with surveys and by reviewing traffic data (Perin, 2004). In the study, it was revealed that many centers are underutilized by students. It was also noted that students who put in six or more visits a semester to the centers are pulling in higher GPAs and experiencing better success in the course (Perin, 2004).

Reviewing the data of your learning centers can help with understanding how to utilize your centers and increase traffic. The learning centers need more developmental students to visit and access their resources. Learning centers should create an environment of comfort and support for students who may find going to a center an intimidating experience. The numbers show that the more often students visit, the more their chances of success increase.

JOB SATISFACTION AND FACULTY EXPECTATIONS

In exploring various studies about workshops being aimed to improve the teachers' views of mathematics and learning mathematics, a common theme can be used to help develop a conceptual framework for the workshops to help community college instructors' mindsets. Improving mindset can help with overall job satisfaction and expectations with teaching. The sudden change in reform for colleges and universities could have some negative effects on instructors, especially if no professional development was attended by that instructor.

A study was done by Carmen Latterell and Janelle Wilson to analyze the metaphors that preservice teachers came up with to describe their feelings about mathematics and learning mathematics. For teaching and life in general, "No one wants to enter a process that is viewed to be unpleasant" (Latterell & Wilson, 2016, p. 291). The study showed that after the teachers created their metaphor and it was categorized, most of the results were viewed as negative (see table 4.1). The most-selected categories were (1) dangerous, difficult, unpleasant, impossible; and (2) endless, expansive process (Latterell & Wilson, 2016, p. 291).

Table 4.1. Results of the Latterell & Wilson Study

Category	Math Is . . .	Learning Math Is . . .
Dangerous, Difficult, Unpleasant, Impossible	22	25
Easy, Pleasant	7	7
Endless, Expansive Process	32	47
Puzzle	16	4
Necessary	13	3
Language	2	5

Source: See Latterell & Wilson, 2016.

After the study, a discussion was held on how important it is to recognize these feelings and how to address them. In order to go into the classroom, teachers need to have a clear understanding of what creates these thoughts and what can be done to improve them so as to not stigmatize the class. To see a technique to address these feelings, an experiment was conducted on another group of teachers.

The third technique involves a study done by Sonja Mohr and Rossella Santagata, which used preservice teacher preparation courses to impact their belief change (Mohr & Santagata, 2015). The study involved video-enhanced mathematics method courses to help prepare preservice teachers for the year. In the study, it is mentioned that teachers' beliefs have four characteristics: "They influence perception, they are dispositions to actions, they are held with differing intensities, and they tend to be context specific" (Mohr & Santagata, 2015, p. 104). Having these characteristics enter a classroom can affect the environment, so the study investigates ways of changing these beliefs.

The preservice teachers attended two 20-week mathematics method courses, meeting three hours a week. The course utilized videos to analyze teaching and the ways students learn. The collection of videos displayed mathematics teaching that was demonstrated to be the most effective at reaching the students. While watching the videos, teachers were asked to reflect by answering a series of questions that helped with analyzing the videos and with discussion.

Before and after the course, the teachers were surveyed using the Integrating Mathematics and Pedagogy (IMAP) survey and, for the sake of time, only four of the seven beliefs were used (Mohr & Santagata, 2015):

Belief 1. Mathematics is a web of interrelated concepts and procedures (and school mathematics should be too). Beliefs about learning or knowing mathematics, or both.

Belief 2. If students learn mathematical concepts before they learn procedures, they are more likely to understand the procedures when they learn them. If they learn the procedures first, they are less likely ever to learn the concepts. Beliefs about children's (students') learning and doing mathematics.

Belief 3. Children can solve problems in novel ways before being taught how to solve such problems. Children in primary grades generally understand more mathematics and have more flexible solution strategies than adults expect.

Belief 4. During interactions related to the learning of mathematics, the teacher should allow the children to do as much of the thinking as possible.

The survey was done before the workshop and during the workshop; the post-survey shows the design and logic of the study in collecting data. The survey creates the source of evidence and measurement to observe the usefulness of this design for researching purposes.

Based on the results from the surveys, there is a significant change in the beliefs of the teachers and their teaching and learning of mathematics. The beliefs before are traditional, fixed mindsets when it comes to teaching, but the results afterward show movement toward a more progressive mindset about mathematics. There was a huge shift in believing that students who learn concepts of mathematics before learning procedures have a better chance at understanding, as well as a huge shift in believing that students can solve problems in novel ways before being taught (see tables 4.2 and 4.3).

Table 4.2. Pre-Survey for Mohr & Santagata Study

	No Evidence	Weak Evidence	Evidence	Strong Evidence	Very Strong Evidence
Belief 1	36%	34%	15%	14%	0%
Belief 2	19%	32%	35%	14%	0%
Belief 3	34%	29%	22%	11%	4%
Belief 4	60%	21%	12%	7%	0%

Source: See Mohr & Santagata, 2015.

Table 4.3. Post-Survey for Mohr & Santagata Study

	No Evidence	Weak Evidence	Evidence	Strong Evidence	Very Strong Evidence
Belief 1	6%	27%	34%	32%	0%
Belief 2	2%	7%	27%	64%	0%
Belief 3	9%	20%	33%	28%	9%
Belief 4	29%	27%	33%	10%	0%

Source: See Mohr & Santagata, 2015.

The fact that so many teachers were able to change their mindset about the ability of their students by simply watching various videos on methods that work is a powerful discovery. Reform is difficult because of the unknown, the change in the norm; and being provided with some concrete examples of pedagogy can assist faculty in reform tensions.

Generalization of this study can be made to reach groups of teachers who are experiencing change. The workshops were designed to view the students in a different light. Students taking developmental courses can be assumed to be in a group that may not make it to college level. One teacher from the University of Texas at Austin was quoted saying, "Developmental Math is where aspirations go to die" (Center for the Study of Social Policy, 2016, p. 2). This view can possibly be changed with workshops like the one in this article, especially since campuses are looking at new reforms like corequisites that are showing significant improvements in developmental education.

CARNEGIE MATH PATHWAYS REFORM LEADS TO FACULTY SUPPORT PROGRAM

An extensive blueprint for starting a professional development for developmental education can be seen with faculty training created in the implementation of the Carnegie Community College Pathways. The 2015 article by Ann Edwards, Carlos Sandoval, and Haley McNamara titled "Designing for Improvement in Professional Development for Community College Development Mathematics Faculty" is extremely enlightening and discusses the ups and down of developmental education, the reform of developmental education, and the necessity of making sure the voices of faculty are heard when it comes to providing professional development to combat the changes.

The article discussed the framework, the reformatting of the program, and the results of the research-based design of the program. Just as with the corequisite model mandate, the pathway reform was a huge change, but professional development was not the priority when the change in education was placed.

In 2009, there was a big reform with mathematics that dramatically increased the college-level success rates for those who were placed in developmental mathematics courses. This reform was labeled Carnegie Math Pathways (Center for the Study of Social Policy, 2016). Because of the magnitude of the reform, there needed to be some type of professional development to help with the implementation of this program.

The creation of the professional development needed to enhance teaching in this new format was called the Faculty Support Program (FSP), a training system for first-time teachers of the Carnegie Community College Pathways (Edwards et al., 2015). The program was research based and designed to make sure it was the best format for the needs of the faculty.

The design of the FSP was based on a set of principles pieced together from research on professional development for the K–12 setting and they are the following:

1. Program structure provides for sustained opportunities for professional learning.
2. Learning activities are job-embedded, supporting emergent problems of practice.

3. Learning activities are context/discipline specific.
4. Learning activities provide opportunities for collaborative reflection.
5. Learning activities are centered around artifacts/representations of classroom practice. (Edwards et al., 2015)

As stated before, the teaching load of a community college instructor can be heavy, so the time needed for professional development must be used valuably. Any type of training taking the teachers away from preparation of their classes needs to be of value. It needs to be training that can directly be used in their day-to-day teaching and interaction with students (Fullan & Miles, 1992). The principles used to create FSP make sure that the creation of the program kept faculty's time and benefits in mind.

These principles ensure that the program is interactive, collaborative, hands-on, job-specific, and overall valuable for the advancement of higher learning educators. These principles can be used to enhance the professional development already offered for corequisite models or even to create new ones that are campus specific.

The FSP principles were created from observation of K–12 education and, after a year, there was a need for redesign. An interesting part of the FSP design was the technique for improvement. The designers used a technique called improvement science to observe the first year of the program. Improvement science "provides a set of tools, approaches, and methodologies for designing and improving systems that are user-centered, responsive to varying and changing conditions, and are structured for continuous learning" (Edwards et al., 2015, p.470).

Improvement science also has principles that back the science behind why this technique works. This tool pointed out some key improvements that were needed to really hear the voices of those attending the FSP, not just a simple "did it work or not" approach. Redesign methods also included tools from Hasso Plattner Institute of Design at Stanford and IDEO, which had five phases of user-centered design process: empathy, define the problem, ideate, prototype, and test (Brown, 2009; Edwards et al., 2015).

This makes redesign meaningful and focused for educational programs; the article shows how the design phases helped with redesign for FSP. This redesign required surveys and interviews to investigate the areas of improvement for the program; and the creators of the program do an intentional job of understanding the key steps in each phase and how they redirect back to improvement for FSP.

Generalization of this study shows the usefulness of taking this design and approach to create a college's professional development to improve faculty relationship with the reform. It creates a blueprint of professional development. To help with understanding how to interpret data (qualitative and quantitative), there is an assessment of the forum workshop for faculty provided, which shows a real example of coding of qualitative data collected (Edwards et al., 2015). Overall, the FSP shows many focuses of good practices in creating valuable training for instructors and aids in major reforms in education.

With the reform of the developmental courses switching to corequisite, the addition of professional development for the instructors of these courses can prove to be beneficial; and it is not too late. As the data from the interviews were analyzed, the resources show that there is a plethora of programs available to help with various issues that arise. One just needs to understand the needs of their campus. Once the issues are revealed, a plan can be put into play. Collaboration among faculty with professional development can help raise the standards of math education. It will put faculty in the position of professor readiness, which in turn helps us prepare our students to be college-ready.

Chapter Five

Conclusion

> For some professionals, embracing and implementing change is natural or even intuitive . . . For them, this process is comfortable. For others, any change can be challenging. . . . Regardless of whether the process of change is natural or challenging, continuous improvement in mathematics instruction is essential to improving student learning.
>
> —Blair 2006, p. 9

WHAT DID WE LEARN?

The need for professional development for instructors of corequisite and other developmental mathematics courses forces professional development providers to realign their approaches and strategies to meet the needs and challenges in our field. The corequisite model has been a significant challenge for instruction within mathematics departments; and to go even further, making it a mandate added another layer of pressure, forcing instructors to adjust to these changes very quickly. Should professional development be a requirement based on the challenges that were found in the interviews?

The data revealed some common trends that could be addressed and some common mistakes that could be avoided altogether with proper and effective training and collaboration. What impact does professional development really have on our corequisite faculty? Data provide those answers. The data shows that the professional development ideas and suggestions provided in chapter 4 do provide improvement on campuses. The research conducted for this book has hopefully provided proof of the validity of training to prepare corequisite faculty, because our students' readiness starts with our readiness as educators. As educators, we should be committed to ongoing professional learning and development to stay current with the needs of students in these developmental classrooms.

Most reforms—whether from the state, district, or at the school level—can generally impact faculty more than we realize, and the way the components of the reform are enacted may require instructors to implement many different pedagogical approaches. There are some institutions that have taken advantage of professional development when it comes to implementing corequisites. Tulsa Community College requires training for any instructor teaching using the new format. It consists of six hours of training with adjunct instructors receiving a stipend of $200 for attending (Tulsa Community College, n.d.).

Complete College America offers corequisite regional meetings and houses webinars on their website. Texas State University has the education institute for their campus, but it is not mandatory. Austin Community College has the Texas Corequisite Project, which is labeled as the statewide professional development offering for colleges and universities. Again, it is not mandatory, but one can get information and attend virtual webinars that are housed on their website.

Unfortunately, the Texas Corequisite Project has no known data reporting its benefits or failures concerning its impact on faculty who teach in the corequisite model since it was mandated in 2017. As can be seen from research presented in the book, the results can be helpful for community college instructors to generate some concrete data demonstrating if the model in their institutions "works" or if there is a need to improve specific components of their new model implementation of corequisites. A group of studies should be conducted on current corequisite professional development models to determine if there is room for improvement and to assist instructors to implement corequisite courses more effectively.

There is a need for professional development, but there is also a need for research data to demonstrate that the professional development that is currently being enacted with corequisites is indeed beneficial for instructors and students. In summary, through this book, there is reflection on the history and the current state of corequisite courses in Texas (chapter 1); on various pedagogical approaches from community colleges across the country in mathematics (chapter 2); on faculty voices on their experience with corequisite and developmental mathematics (chapter 3); and, lastly, on professional development ideas based on various themes that emerged from interviews (chapter 4).

This book is simply a starting point to enlighten others concerning corequisites and the start of conversations about making sure corequisites are taught by instructors who are teaching to the best of their abilities.

CHAPTER 1 SUMMARY

Chapter 1 of this book is an overview of mathematics corequisites at community colleges. This chapter is a literature review of the prior and current research available about corequisite and developmental mathematics education. As an overview of the policies affecting developmental education and the policies that led up to the mandate of corequisites at institutions of higher education, this chapter noted the significant changes that have occurred over the years. In this chapter, the foundation was set for why further emphasis should be placed on providing training concerning the ramifications of these changes. The research was reviewed supporting these changes, including previous negative research against these changes. Also included was a discussion of the significance in Texas of its statewide mandate and the data that led to that mandate. Paired with history and present policies, chapter 1 provided the introduction to developmental education and the shift to corequisite education.

CHAPTER 2 SUMMARY

As was previously mentioned, there are colleges and universities across the United States that have incorporated some types of alternatives to the traditional developmental courses offered for mathematics. Among those alternatives are different forms of accelerated programs, including corequisites that are aimed at motivating underprepared to students to enroll in college-level courses. This chapter was dedicated to sharing researched-based effective teaching and learning pedagogies for improving student learning for those enrolled in developmental mathematics classes, mostly at the higher education level. Many colleges and universities have developed their own minor variations of corequisites.

It is worth noting these minor modifications to determine which pedagogies are successful and which cause more complications when implementing these types of courses. Colleges such as LaGuardia Community College have a supplemental instructor (SI) program, called academic peer instruction, which is being used to support the corequisite model on their campuses (Jaafar et al., 2021). The Virginia Community College System has initiatives that include multiple measures and "direct enrollment" to assist in increasing successes for underprepared students (Emblom-Callahan et al., 2019). Ivy Tech Community College has shown increased student success since implementing their corequisite program that was based on the new Mathways project, which aligns content to career aspirations (Royer & Baker, 2018).

Examining different studies involving various campuses will help in creating a collection of learning pedagogies which have been proven to show success in improving student achievement. Specifically, locating studies that investigate the pedagogy for teaching college-level courses paired with developmental courses was the ultimate goal of this chapter.

CHAPTER 3 SUMMARY

In chapter 3, the reader is allowed to gain an understanding of the current corequisite and developmental programs at various community colleges by listening to the faculty voices at those same campuses. This chapter was dedicated to interviewing unique participants and analyzing the author's data collected specifically for this book about how corequisites are currently being implemented at various institutions in Texas.

The framework for conducting this research was based on a qualitative study done on job satisfaction, "A Qualitative Method for Assessing Faculty Satisfaction" (Ambrose et al., 2005). In the quest to collect feedback and data from faculty about their job satisfaction, this group opted to conduct semi-structured interviews in lieu of predefined surveys (Ambrose et al., 2005). When collecting data from the faculty who teach corequisites, the goal was to listen to the voices of the participants; since those voices might be limited by the predefined surveys that were conducted, the interviews were included to help shed light on the real thoughts of the mandate of corequisites.

Coding revealed any overall universal patterns that overlapped between campuses, which was evident in many instances. The interviews even provided recognition for some optimistic views and perspectives that might not have been realized before the conversations.

The research employed for this book involved interviews conducted with faculty from three two-year institutions. The syllabi were pulled from each campus to compare structure, grading systems, modalities, and resources available for corequisite mathematics. The syllabus from each person interviewed was pulled to match the syllabus with the interviewee for triangulation. Two faculty members from each campus were interviewed using the following questions:

1. How long have corequisites been on your campus?
2. Explain the model of the college algebra corequisites. Provide the most successful component of corequisites on your campus.
3. Provide the hardest challenge of corequisites on your campus.
4. Does your campus offer professional development to help with corequisite teaching?
5. If yes, what components of the professional development are most useful?

Having interviews on multiple campuses provided opportunities to examine validity of the developed themes and categories through triangulations by analyzing the research question from multiple perspectives (Creamer, 2018). The data collected highlighted interesting perspectives; for example, there was a close relation to the format of the course versus job satisfaction.

Faculty at colleges that had a corequisite format which shared teachers had more positive experiences in teaching. There seems to be a benefit in having a colleague to help balance the course and understand best practices, which further reveals the significance professional development could bring to corequisite instructors. Overall, chapter 3 leaves the reader with a feel for how these courses are affecting the professors.

CHAPTER 4 SUMMARY

The overall goal of the book is to provide a starting place for colleges to develop their own professional development based on evidence provided in literature review (chapters 1 and 2) and evidence emerging from the research conducted (chapter 3) by interviewing the participants. Chapter 4 shared practices and workshop activities that can be conducted on campuses that are looking to improve their corequisite mathematics or developmental mathematics classes.

The techniques and resources provided in this section are a direct reflection of combining the analyzed data from interviews and syllabi reviews, which highlighted the successes and areas for improvement when it comes to running corequisites on each campus. Throughout this chapter, I demonstrated how to make the professional development specific to the needs of individual campuses.

The chapter also shed light on professional development that is available, such as the Faculty Support Program. This professional development was created to enhance teaching in the Carnegie Pathways format; called FSP, it is a professional development system for first-time teachers of the Carnegie Community College Pathways (Edwards et al., 2015). The program was research based and reformed after a year to ensure it was designed in the best format to meet the needs of the faculty.

Another successful professional development involved a study done by Mohr and Santagata (2015) using preservice teacher preparation courses to impact their belief change. Their study involved video-enhanced mathematics method courses to help prepare preservice teachers for the year. There are resources and techniques that are available for campuses to enhance the way developmental education is presented to students. This chapter is a starting point on proactive practice for faculty members to consider.

What is next? A leading next step would be a more detailed survey to understand the feelings of instructors of the corequisite model, just as the interviews revealed. The interviews were a small sample size of faculty members. A larger selection of faculty would be a great next step to determine if the themes are consistent. There is an interesting component of how faculty feel about the classes they teach, and that component can be linked to job satisfaction. Exploring studies that discuss views on teaching and learning, there was a connection with teachers' job satisfaction and the factors that impact it.

Job satisfaction is an important concept for society to consider, because people spend such a large portion of their lives working. The way one feels about work can affect the entire mood of that person. We want our educators to feel good about what they do and what they teach. The mental aspect in an educator's career is important across multiple studies. When major changes happen in education, even on a collegiate level, attention needs to be pointed to the mental aspect of that change. A mandate like corequisites, when coupled with limited coping strategies to adapt to these major changes, can impact educators and, unfortunately, in a negative way. Factors that link directly to their overall career and teaching are topics that an instructor values when tied to job satisfaction.

A useful framework for the next study should mimic a thematic analysis done by Lehner and Kaniskas in 2012. Their study was a thematic meta-analysis in which qualitative data were collected from 77 different studies to compare common themes and develop new ones. The studies included had to have various qualifications to be considered for the meta-analysis. Data analysis consisted of reviewing papers individually and then conducting a cross-paper comparison within the different categories (Lehner & Kaniskas, 2012).

After reviewing the papers individually and collecting data from each study, they were able to develop codes to categorize them under common themes. This framework is the approach taken in the study of job satisfaction. Upon reading the results of the articles, the themes, codes, and factors of job satisfaction were collected for this thematic analysis. From the search, 11 articles were collected to analyze their results. The collections of themes, codes, and factors were then analyzed in order to group them into major categories that were common across all the articles collected.

The new themes that emerged will help us link some factors that relate to satisfaction in teaching new formats like corequisites. The results can help with addressing some significant questions about teaching when new reforms are introduced to educators.

To have a base for observing the programs and articles, the AERA standards can be the framework for the observations. The eight standards are the following: problem formulation; design and logic of the study; sources of evidence; measurement and classification; analysis and interpretation; generalization; ethics in reporting; and lastly, title, abstract, and heading (Educational Researcher, 2006).

The standards are key to having a thorough reporting of the research for the American Educational Research Association (AERA) and, most likely, good for researching within any organization and its readers. AERA is the baseline for researchers. The standards help "assist researchers in the preparation of manuscripts that report such work, editors and reviewers in the consideration of these manuscripts for publication, and readers in learning from building upon such publications" (Educational Researcher, 2006, p. 33).

Ultimately these standards are important in the investigation of the programs for this paper. Now that the key terms and phrases are explained, program observations can begin. These AERA standards are a creditable way to measure the level of influence of an adapted program.

BLUEPRINT TO IMPROVE DEVELOPMENTAL MATH LEARNING

To end this manuscript, there is a blueprint that combines the research and the needs that campuses may experience. The first step in making any change is questioning. The only way to learn is by questioning. Ask the faculty their opinions and feelings about what they face in the classroom. The biggest mistake is to make assumptions about the feelings of the faculty who face the students every day and work closely with them. Figure 5.1 shows a path to follow in creating your professional development to improve math learning. A sample survey is provided in figure 5.2. The survey is simple in order to give the faculty freedom to expand on their thoughts.

Once the questions are asked and answers are collected, common themes will emerge, and you will become the researcher as you analyze this information. Suggestions for the data collection include the following: use the most successful answer to help with ideas for professional development and use the most challenging to see if any answers fall into the themes found in this book. If the most challenging answers do not fall into one of the eight themes found here, create a new theme to add to the list. Again, figure 5.1 is about the path to take and figure 5.2 gives a sample of a survey used to request information from faculty about developmental and corequisite learning.

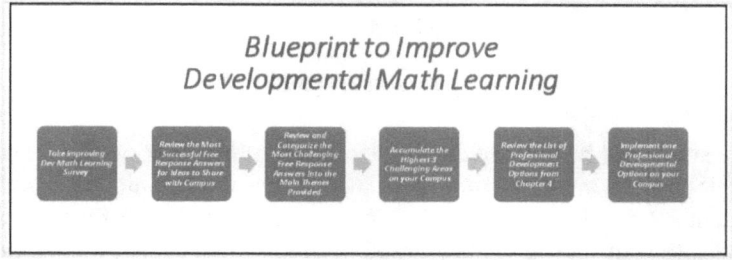

Note: The above blueprint is a guided direction to start your campus learning improvements.

Figure 5.1.   Blueprint to Improve Developmental Math Learning

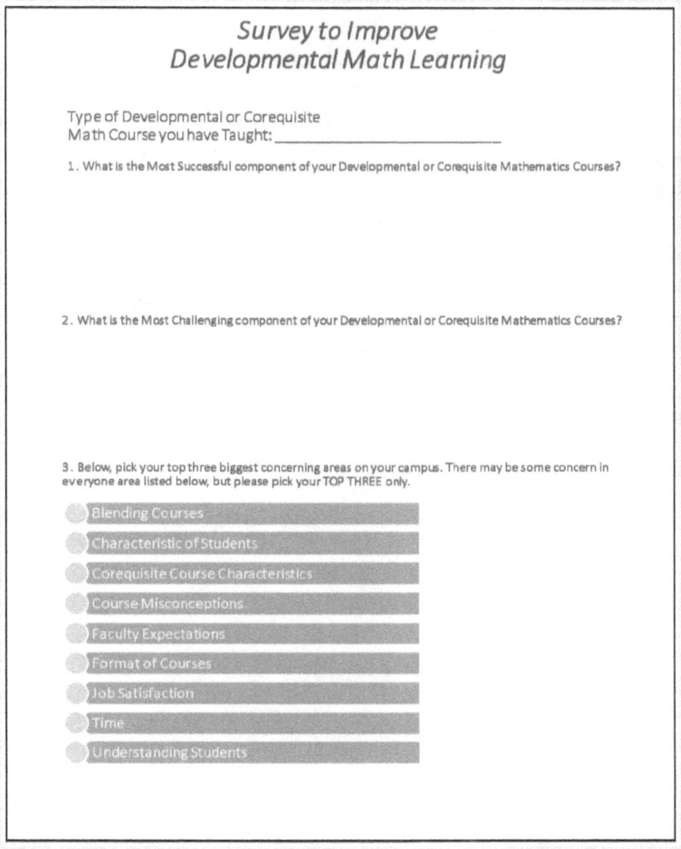

Survey to Improve Developmental Math Learning

Type of Developmental or Corequisite
Math Course you have Taught: _____

1. What is the Most Successful component of your Developmental or Corequisite Mathematics Courses?

2. What is the Most Challenging component of your Developmental or Corequisite Mathematics Courses?

3. Below, pick your top three biggest concerning areas on your campus. There may be some concern in everyone area listed below, but please pick your TOP THREE only.

- Blending Courses
- Characteristic of Students
- Corequisite Course Characteristics
- Course Misconceptions
- Faculty Expectations
- Format of Courses
- Job Satisfaction
- Time
- Understanding Students

Note: The above figure is a simple survey to start the professional development analysis. The results can be used to refer back to chapter 4 to get professional development ideas.

Figure 5.2.   Survey to Improve Developmental Math Learning

REFLECTION

This study has been a welcoming experience in exploring the field of developmental mathematics with such a great and unique group of students. Adult learners all over the world are faced with fundamental studies to start their college journey in becoming leaders of our future. We have a role as educators in those fundamental courses, and that is to see them through it. The wonderful thing that sparks such a huge interest in developmental education is that we, the faculty, are not just there to teach mathematics. We have a role in helping people become successful college students; we have a role in helping them navigate their lives while in college. It is not just about teaching mathematics; it is about developing a person into a scholar for any course.

I trust this study has opened your eyes to the fact that even though we have changes in our math departments, we owe it to our students to make sure each and every one of us are becoming the best instructors we can be for those changed math courses. It may be overwhelming, it may be a lot to handle, but how we approach challenges makes a difference in how we react and handle it. We need to approach developmental education with as many tools as we can amass so that we can be champions for our students in successful and effective corequisites classes. We want to push as many students as we can to become college-ready because we will take a proactive approach in making sure we have professors ready and able to teach effectively based on hard data.

References

60x30TX. (2018, January 25). FAQs: HB 2223 implementation. Texas Higher Education Coordinating Board. https://reportcenter.highered.texas.gov/agency-publication/miscellaneous/faq-hb-2223-tsi-de/

Adams, T. (n.d.) Department of Mathematics. Community College of Denver. https://www.ccd.edu/program/mathematics

Ambrose, S., Huston, T., & Norman, M. (2005). A qualitative method for assessing faculty satisfaction. *Research in Higher Education, 46*(7), 803–30. https://doi.org/10.1007/s11162-004-6226-6

Austin Community College (n.d.) Texas Corequisite Project. Texas Higher Education Coordinating Board. https://instruction.austincc.edu/txcoreqs/

Bishop, T. J., Martirosyan, N., Saxon, D. P., & Lane, F. (2018). Delivery method: Does it matter? A study of the North Carolina developmental mathematics redesign. *Community College Journal of Research and Practice, 42*(10), 712–23. https://doi.org/10.1080/10668926.2017.1355281

Blair, R. (2006). Beyond crossroads: Implementing mathematics standards in the first two years of college. *College Mathematics Journal, 39*(4), 324.

Booth, E. A., Capraro, M. M., Capraro, R. M., Chaudhuri, N. B., Dyer, J. A., & Marchbanks III, M. P. (2014). Innovative developmental education programs: A Texas model. *Journal of Developmental Education, 38*(1), 2–18. https://files.eric.ed.gov/fulltext/EJ1071605.pdf

Borko, H., Jacobs, J., Seago, N., & Mangram, C. (2014). Facilitating video-based professional development: Planning and orchestrating productive discussions. In Y. Li, E. A. Silver, & S. Li (Eds.), *Transforming mathematics instruction: Advances in mathematics education* (pp. 259–81). Springer, Cham. https://doi.org/10.1007/978-3-319-04993-9_16

Brown, T. (2009). *Change by design: How design thinking transforms organizations and inspire innovation.* Harper Business.

Burrows, A. (2015). Partnerships: A systemic study of two professional developments with university faculty and K–12 teachers of science, technology, engineering, and mathematics. *Problems of Education in the 21st Century, 65,* 28–38. https://doi.org/10.33225/pec/15.65.28

Campos, E. (2020, September 29). What is professional development? Study. com. https://study.com/academy/popular/what-is-professional-development.html

Center for the Study of Social Policy. (2016). *Carnegie math pathways: Case study, DC.* https://cssp.org/wp-content/uploads/2018/08/Carnegie-Math-Pathways.pdf

Charmaz, K. 2000. Constructionism and the grounded theory method. In J. A. Holstein & J. F Gubrium (Eds.), *Handbook of constructionist research* (2nd ed., pp. 397–412). New York: Guilford.

Chen, X. & Simone, S. (2016). *Remedial coursetaking at U.S. public 2- and 4-year institutions: Scope, experiences, and outcomes.* Statistical analysis report. National Center for Education Statistics.

Complete College America. (2016). *New rules: Policies to strengthen and scale the game changers.* Complete College America.

Cox, R. (2015). "You've got to learn the rules": A classroom-level look at low pass rates in developmental math. *Community College Review, 43*(3), 264–86.

Creamer, E. (2018). *An introduction to fully integrated mixed methods research.* SAGE. https://dx.doi.org/10.4135/9781071802823

DeJarnette, A. F., & Hord, C. (2020). Pre-service teachers' patterns of questioning while tutoring students with learning disabilities in algebra 1. In A. I. Sacristán, J. C. Cortés-Zavala, & P. M. Ruiz-Arias (Eds.), *Mathematics education across cultures: Proceedings of the 42nd meeting of the North American chapter of the International Group for the Psychology of Mathematics Education, Mexico.* https://doi.org/10.51272/pmena.42.2020-250

Eades, C., & Moore, W. M. (2007). Ideas in practice: Strategic note-taking in developmental mathematics. *Journal of Developmental Education, 31*(2), 18–26.

Edgecombe, N. (2011). Accelerating the academic achievement of student referred to developmental education. *Community College Research Center*, 1–45.

Educational Researcher. (2006). Standards for reporting on empirical social science research in AERA publications. American Educational Research Association. *Educational Researcher, 35*(6), 33–40.

Edwards, A. R., Sandoval, C., & McNamara, H. (2015). Designing for improvement in professional development for community college developmental mathematics faculty. *Journal of Teacher Education, 66*(5), 466–81. https://doi.org/10.1177/0022487115602313

Ellerbe, L. W. (2015). Faculty expectations for student success in community college developmental math. *Community College Journal, 39*(4), 392–96.

Emblom-Callahan, M., Burgess-Palm, N., Davis, S., Decker, A., Diritto, H., Dix, S., Parker, C., & Styles, E. (2019). Accelerating student success: The case for corequisite instruction. *Inquiry: The Journal of the Virginia Community Colleges, 22*(1). https://commons.vccs.edu/inquiry/vol22/iss1/12/

Frost, J. H., Coomes, J., & Lindeblad, K. K. (2009). Collaborating to improve students' transitions from high school mathematics to college: Characteristics and outcomes of a cross-sector professional developmental project. *NASSP Bulletin, 93*(4), 227–40.

Fullan, M. G., & Miles, M. B. (1992). Getting reform right: What works and what doesn't. *Phi Delta Kappa International, 73*(10), 744–52.

Geltner, P., & Logan, R. (2001, April 20). *The influence of term length on student success*. Research report ERIC No. E455858. Santa Monica College, CA.

Gerver, R. (2018). *Write On! Math: Note talking strategies that increase understanding and achievement*. Information Age Publishing.

Goudas, A. M. (2017). *The corequisite reform movement: A higher education bait and switch*. Community college data. http://communitycollegedata.com/articles/the-corequisite-reform-movement/

Heick, T. (2012). *The Art and Science of Reading in the 21st Century: Bloom's Taxonomy, ela, Experienced Teacher*. www.teachthought.com

Hern, K. (2012). Acceleration across California: Shorter pathways in developmental English and math. *Change: The Magazine of Higher Learning, 44*(3), 60–68.

House Research Organization. (2017, May 5). *Bill analyses*. Texas House of Representatives. https://hro.house.texas.gov/BillAnalysis.aspx

Jaafar, R., Cornelius, A. M., Zaritsky, J., Evering, J., & Pal, A. (2021). Supplemental instruction in corequisite mathematics courses: Results and challenges at an urban community college. *MathAMATYC Educator, 12*(2), 36–68.

Jaggars, S. S., Hodara, M., Cho, S.-W., & Xu, D. (2014). Three accelerated developmental education programs: Features, student outcomes, and implications. *Community College Review, 43*(1), 3–26. https://doi.org/10.1177/0091552114551752

Joyce, B., Weil, M., & Calhoun, E. (2015). *Models of teaching* (9th edition). Upper Saddle River, NJ: Pearson.

Khoule, A., Pacht, M., Schwartz, J. W., & van Slyck, P. (2015). Enhancing faculty pedagogy and student outcomes in developmental math and English through an online community of practice. *Research & Teaching in Developmental Education, 32*(1), 39–49.

Koch, B., Slate, J. R., & Moore, G. (2012). Perceptions of students in developmental classes. *Community College Enterprise, 18*(2), 62–82. https://eric.ed.gov/?id=EJ1001036

Latterell, C. M., & Wilson, J. L. (2016). Math is like a lion hunting a sleeping gazelle: Preservice elementary teachers' metaphors of mathematics. *European Journal of Science and Mathematics Education, 4*(3), 283–92. https://doi.org/10.30935/scimath/9470

Lehner, O. M., & Kaniskas, J. (2012). Opportunity recognition in social entrepreneurship: A thematic meta analysis. *Journal of Entrepreneurship, 21*(1). https://doi.org/10.1177/097135571102100102

Logue, A. W., Douglas, D., & Watanabe-Rose, M. (2019). Corequisite mathematics remediation: Results over time and in different contexts. *Educational Evaluation and Policy Analysis, 41*(3), 294–315. https://doi.org/10.3102/0162373719848777

Meiselman, A. Y., & Schudde, L. (2020). *The impact of corequisite math on community college student outcomes: Evidence from Texas.* Texas Scholar Works. University of Texas at Austin.

Mellow, G. O., Woolis, D. D., Klages-Bombich, M., & Restler, S. (2015). *Taking college teaching seriously: Pedagogy matters!: Fostering student success through faculty-centered practice improvement.* Stylus.

Mohr, S., & Santagata, R. (2015). Changes in pre-service teachers' beliefs about mathematics teaching and learning during teacher preparation and effects of video-enhanced analysis of practice. *Orbis Scholae, 9*(2), 103–17. https://doi.org/10.14712/23363177.2015.82

Morgan, K., & Morales-Vale, S. (2019). *College readiness.* Texas Higher Education Coordinating Board.

Ngo, F. (2019). Fractions in college: How basic math remediation impacts community college students. *Research in Higher Education, 60*(4), 485–520.

Perin, D. (2004). Remediation beyond developmental education: The use of learning assistance centers to increase academic preparedness in community colleges. *Community College Journal of Research and Practice, 28*(7), 559–582. https://doi.org/10.1080/10668920490467224

Ran, F. X., & Lin, Y. (2019). The effects of corequisite remediation: Evidence from a statewide reform in Tennessee. CCRC working paper No. 115. Community College Research Center (CCRC), Teachers College, Columbia University.

Royer, D. W., & Baker, R. D. (2018). Student success in developmental math education: Connecting the content at Ivy Tech Community College. *New Directions for Community Colleges, 2018*(182), 31–38. https://doi.org/10.1002/cc.20299

Scott-Clayton, J., et al. (2014). Improving the targeting of treatment: Evidence from college remediation. *Educational Evaluation and Policy Analysis*, *36*(3), 371–93. https://www.nber.org/system/files/working_papers/w18457/w18457.pdf

Scrivener, S. Gupta, H., Weiss, M. J., Cohen, B., Cormier, M. S., & Brathwaite, J. (2018). *Becoming college-ready: Early findings from a CUNY start evaluation*. Evaluative report ERIC No. ED586380. MDRC; Community College Research Center, Columbia University.

Silver, E. A., & Suh, H. (2014) Professional development for secondary school mathematics teachers using student work: Some challenges and promising possibilities. In Y. Li, E. Silver, & S. Li. (Eds.), *Transforming mathematics instruction: Multiple approaches and practices* (pp. 283–309). Springer. https://doi.org/10.1007/978-3-319-04993-9_17

Smith, A. A. (2017, July 12). Texas requires credit-bearing remediation. *Inside Higher Ed*. https://www.insidehighered.com/news/2017/07/12/texas-legislature-requires-colleges-use-popular-reform-approach-remedial-education

———. (2016, February 23). Experiencing development education. *Inside Higher Ed*. https://www.insidehighered.com/news/2016/02/23/broad-study-community-college-students-who-take-developmental-education-courses

Squires, J., Faulkner, J., & Hite, C. (2009). Do the math: Course redesign's impact on learning and scheduling. *Community College Journal of Research and Practice*, *33*(11), 883–86.

Texas Higher Education Coordinating Board (THECB). (2019, July). *60x30TX 2019 progress report*. http://reportcenter.highered.texas.gov/reports/data/60x30tx-progress-report-july-2019/

Texas Success Center. (2020, August 17). *Texas corequisite project webinar: Achieving 100% corequisites*. [Video]. YouTube. https://youtu.be/k9fHuMjpNEM

Tulsa Community College. (n.d.). Tulsa community college redesign: Corequisites and mathways. Dana Center Mathematics Pathways. https://dcmathpathways.org/file-download/download/public/698

Weiss, M. J., & Headlam, C. (2019). A randomized controlled trail of a modularized, computer-assisted, self-paced approach to developmental math. *Journal of Research on Educational Effectiveness*, *12*(3), 484–513. https://doi.org/10.1080/19345747.2019.1631419

Yang, Y. (2014). How classroom instruction was improved in a teaching research group: A case study from Shanghai. In Y. Li, E. Silver, & S. Li. (Eds.), *Transforming mathematics instruction: Multiple approaches and practices* (pp. 355–81). Springer. https://doi.org/10.1007/978-3-319-04993-9_20

About the Author and Contributors

ABOUT THE AUTHOR

Dr. Ajai Cribbs Simmons is a mathematics professor at a community college in Texas, where her passion is improving developmental and corequisite math courses. She has been teaching since 2009, starting her employment in California. Simmons is also an Instructional Skills Workshop (ISW) facilitator; this program helps employees fulfill their capstones for the Higher Education Teaching Institute (HETI) program. The goal of ISW is to step away from the details of content and enhance the teaching effectiveness in classrooms. The idea is that "we are all masters of our content, but improvements can be made in our instructional skills." From 2009–2012, Simmons facilitated an 80-hour workshop called Intel Math for K–8 teachers to help deepen their understanding of math to strengthen their ability to teach their students.

Dr. Simmons holds a BS in Mathematics from University of Alabama in Huntsville and an MS in Applied Mathematics from University of Alabama at Birmingham. She earned her doctorate in curriculum and instruction, specializing in mathematics education, from Texas A&M University. Her focus was on professional development for math instructors who teach at community colleges. She continues her career in education; she is also a wife and the mother of two.

ABOUT THE CONTRIBUTORS

Dr. Tasha Bennett spent over 21 years in financial services, specifically in property and casualty insurance, and her passion for people development can be traced back to those experiences. She was able to help people positively impact business through training, which is why human resource development was an optimal career choice for her. Currently, she is a learning and development supervisor for Marathon Oil.

Learning is one of the most important aspects of our lives. As an educator, Dr. Bennett enjoys using her experiences, skills, and adaptability to help create a learner-centric environment for her students. She has over 13 years of experience teaching in synchronous and asynchronous environments, with more than three years at the University of Houston. An important aspect of Dr. Bennett's life is family. In addition to family, she is very active in a variety of civic and cultural organizations.

Dr. Bennett has degrees in management and organizational leadership from University of Phoenix; an MA in Industrial Organizational Psychology from Louisiana Tech University; and a BS in Business Administration from University of Arkansas at Pine Bluff.

Brittany Garcia-Pi is a PhD student in the architecture department at Texas A&M University, majoring in Human Computer Interaction. Her research centers around the design and user experience of virtual reality applications for educational training and collaboration. When Brittany is not conducting research, she spends time watching documentaries with her husband and their dog, Ellie.

www.ingramcontent.com/pod-product-compliance
Lightning Source LLC
Chambersburg PA
CBHW032030230426
43671CB00005B/269